DATE DUE
AUG
W9-AVT-189

RARE BREEDS

RARE BREEDS

TERRY BRIDGE

CHARTWELL
BOOKS, INC.

Published in 2010 by
CHARTWELL BOOKS, INC.
A division of BOOK SALES, INC.
276 Fifth Avenue Suite 206
New York
NY 10001
USA

The Red House
84 High Street
Buntingford, Hertfordshire
SG9 9AJ, UK

For all editorial enquiries, please contact
Regency House Publishing at
www.regencyhousepublishing.com

ISBN-13: 978-0-7858-2646-0

ISBN-10: 0-7858-2646-7

Printed in China

CONTENTS

INTRODUCTION

Livestock refers to one or more domesticated animals, raised in an agricultural setting to produce commodities such as food, fibre, or labour. Livestock are raised for subsistence or for profit. Raising animals (animal husbandry) is an important component of modern agriculture, and has been practised in many cultures since the transition was made to farming by hunter-gatherers.

Selecting animals for breeding with superior growth-rate, egg-, meat-, milk, or wool-production, or with other desirable traits, has revolutionized agricultural livestock production throughout the world. The scientific theory of animal breeding incorporates population genetics, quantitative genetics and statistics, and is based on the pioneering work of Sewall Wright, Jay Lush, and Charles Henderson.

'Breeding stock' is a term used to describe a group of animals destined for planned breeding. When the intention is to breed animals, people look for certain valuable traits in purebreds, or may decide to engage in crossbreeding to produce a new type of stock with different, and presumably superior, abilities in a

ABOVE: Alpine goats are hardy, resilient animals that thrive in any climate while maintaining good health and excellent production.

OPPOSITE: The Aberdeen Angus was developed in the early 19th century from the polled and predominantly black cattle of north-east Scotland.

given area of performance. When breeding swine, for example, the breeding stock must be sound, fast-growing, muscular, lean, and reproductively efficient, while the subjective selection of breeding stock in horses has led to the creation of many horse breeds with particular performance traits.

Mating together animals of the same breed is referred to as purebred breeding. Compared with the practice of mating animals of different breeds, purebred breeding aims to establish and maintain stable traits that the animals will pass on to the next generation. By 'breeding the best to the best', using a certain degree of inbreeding, considerable culling, and selection for 'superior' qualities, it is possible to develop a bloodline or breed that is, in certain respects, superior to the original base stock. Such animals can be entered in a breed registry, also known as a stud book or register, and which is an official list of animals within a specific breed whose parents are known. Animals are usually registered by their breeders when they are still young. The terms stud book and register also provide lists of male animals 'standing at stud', i.e., those animals actively engaged in breeding, as opposed to every known specimen of that breed.

Vietnamese Potbelly pigs are largely kept as pets. Because they are in the same species as ordinary farmyard pigs and wild boars, they are capable of interbreeding with them.

A breed standard, also called a bench standard in animal husbandry, consists of guidelines to be followed to ensure that the animals produced by a breeder or breeding facility conform to the specifics of the breed. Breed standards are devised by breed associations or breed clubs, not by individuals, and are intended to reflect the use or purpose of the species and breed of animal. Breed standards help define the ideal animal of a breed and provide goals for breeders to strive for in improving stock.

In essence, a breed standard is a blueprint for an animal fit for the function for which it was bred, i.e., herding, tracking, etc. Breed standards are not scientific documents, and may vary from association to association and from country to country, even for the same species and breed. There is no one format across all species, and breed standards can change and are updated over time.

Breed standards cover the externally observable qualities of the animal, such as appearance, movement and temperament. It may include a history of the breed, a narrative description, and details of the ideal externally observable structure and behaviour desirable for

the breed. Certain deviations from the standard are considered to be faults, a large degree of deviation from the standard as an excess of faults, while certain defined major faults may indicate that the animal should not be bred, even though its fitness for other uses may not be impeded. An animal that closely matches (conforms to) the breed standard for its species and breed is said to have good conformation. In cattle, breed standards allow for comparisons to be made and advice to be given on the selection of the best breeds to raise.

A rare breed is defined as a breed of livestock that is not common in modern agriculture, though it may have been in the past. Various national and international organizations, such as the United Nations Food and Agriculture Organization, the American Livestock Breeds Conservancy, and the Rare Breeds Survival Trust of the United Kingdom, each define the exact parameters by which rare breeds are defined. Many breeds that qualify as rare by these standards may only have a few thousand or even a few hundred breeding individuals existing in the world today. These organizations pursue conservation of heritage livestock for their unique traits, which may contribute to genetic diversity among animals important for human food supplies and economies, as well as general biodiversity and improvements in animal husbandry.

OPPOSITE: The head of the Nubian or Anglo-Nubian goat is its distinctive breed characteristic, the profile between eyes and muzzle being strongly convex.

RIGHT: The Angora goat is an ancient breed with records of its hair being used for clothing dating back to the 14th century BC. Its valuable hair is widely used to this day and is known as mohair.

CHAPTER ONE
CATTLE

AMERICAN

The American is a breed of cattle native to the United States and known for its heritage as an American bison hybrid. It was developed in the 1950s by a New Mexico rancher looking for beef cattle which could survive on poor fodder in the arid south-west.

Today, the American is one of only a few pure breeds with any known bison genetics, the better-known breed being the Beefalo. Art Jones, the original breeder, began by crossing Hereford, Shorthorn and Charolais, and later added extensive crosses with Brahman cattle and bison. Today, all members of this rare breed display the genetic marker for bison ancestry. American cattle vary in appearance, though nearly all display the floppy Brahman ears, slight humps, and many have horns.

ANCIENT WHITE PARK

A rare breed of horned cattle with ancient herds preserved in Great Britain. The White Park is white except for its points (ears, nose, eye area, feet and hooves, and teats); the points are usually black but may also be red. The tongue should be pink with a dark underside. This is a medium-sized bovine, the mature bulls weighing in at about 2,100lbs (950kg) and cows coming in at about 1,400lbs (635kg).

It bears a resemblance to the Longhorn, in that the horned White

LEFT & OPPOSITE: The White Park is an ancient breed originating in the UK. It is an amalgam of several old British breeds.

Park of today is the product of the mixing of other old British breeds as described thus: 'The White Park is really a mixture of several breeds and includes blood from Longhorn (English, not Texas), Welsh Black and Scottish Highland. This is because it (White Park) originated when several ancient herds of white cattle were merged and in all of these (unlike the Chillingham) there had been crossing with other breeds.'

While the White Park is now one of the breeds bred and selected for beef production, historically it served a dual purpose and was also valued for its milk. It is a breed well-suited to non-intensive production.

Both horned and polled white cattle with red points are mentioned in the ancient Irish epic, 'The Cattle Raid of Cooley' or 'The Táin', that survives in revised editions dating back to at least the 12th century, and which preserves stories from perhaps 1,000 years earlier. Breeders of White Parks refer to the white-horned Bull of Connaught, most often mentioned in the epic, which is not distinguished as having red ears; in prelude and subsequent sections, however, many references to hornless milk-white heifers and cow herds with red ears are made, and always in the context of domesticity rather than wildness. The 'wild' aspect of the breed, now known as the White Park, has its roots in the isolated Chillingham herd, while other horned herds of White Park cattle are, and have been, quite domesticated.

Cattle of a similar type are also mentioned in Welsh laws, made in Deheubarth by a series of rulers from AD 856 to 1197. A herd at Dynefor (Dinefwr), Wales, dates from this time. Herds at Chartley and Chillingham in England and Cadzow in Scotland date to the mid-13th century at a time when the White Park were 'emparked', i.e., enclosed in hunting chases. There were more than a dozen White Park herds in Britain in the early 1800s, but most had been exterminated by the turn of the following century.

ABOVE: A prize Black Angus bull.

OPPOSITE: Black and Red Angus cattle.

There were only five North American herds in 1995, and with a breeding population of fewer than 50 animals, American Livestock Breeds Conservancy now regards this animal's status as critical. White Park in the United States have been DNA-tested to determine a breeding programme to ensure their survival. Besides Canada, the US and the UK, the breed also appears in Australia, Germany and Denmark, but its total numbers equal only approximately 500 purebred females, plus bulls and young stock They are mostly to be found on rare breed farms, such as at Wimpole Home Farm in Cambridgeshire, England, Appleton Farms in Massachusetts, or the B Bar Ranch in Montana, USA.

ANGUS

Angus is a term that refers to two Scottish breeds of cattle which are naturally polled; that is, they don't have horns. Angus is one of the preferred breeds for beef, especially in the United States. The two breeds of Angus cattle are: the Black Angus, which refers to the predominant colour among the original Scottish

Aberdeen Angus population. If a colour is not specified when referring to Angus cattle, it is presumed to be black; and the Red Angus, a breed resulting from the selection of red individuals from the Angus population, which has always had both red and black individuals.

Aberdeen Angus is the original name of the breed, which was developed in Scotland from cattle native to the counties of Aberdeenshire and Angus. The original name is still in use in the United Kingdom, Europe and South America.

Angus cattle are naturally polled and solid black or red, although the udder may be white. Black Angus is the most popular beef variety in the United States, with 324,266 animals hving been registered in 2005.

For some time before the 1800s, the hornless cattle of Aberdeenshire and Angus were called Angus doddies. Hugh Watson can be considered the founder of the breed, being instrumental in selecting the best black, polled animals for his herd. His favourite bull was Old

Jock, born in 1842 and sired by Grey-Breasted Jock. Old Jock was given the number one in the Scottish herd book, when it was founded. Another of Watson's notable animals was a cow, Old Granny, born in 1824 and said to have lived to be 35 years of age, during which time 29 calves were produced. The pedigrees of the vast majority of Angus cattle alive today can be traced back to these two animals.

On 17 May 1873, George Grant brought four Angus bulls back to Victoria, Kansas. He took them to the fair in Kansas City where they were the subject of conjecture, at a time when Shorthorns and Longhorns were the norm. The black hornless animals were often called 'freaks' by those who saw them. The bulls were used only in crossbreeding and have no registered progeny today. But their offspring made a favourable impression on the cattlemen of the time and soon more Angus cattle were being imported from Scotland to form purebred herds.

On 21 November 1883, the American Aberdeen Angus Association was founded in Chicago, Illinois, but was shortened in the 1950s to the American Angus Association. The association's first herd book was published on 1 March 1885, at which time both red and black animals were registered without distinction. In 1917, however, the association barred red and differently coloured animals in an effort to promote a solid black breed. Red Angus cattle occur as the result of a recessive gene, and breeders collecting red cattle from black herds began the Red Angus Association of America in 1954. Other countries, such as the United Kingdom and Canada, still register both colours in the same herd book.

OPPOSITE: Black Angus cattle grazing.

BELOW: The Red Angus is the result of a naturally-occurring recessive gene.

ANKOLE-WATUSI

Also known as the Ankole or the Watusi, this is a breed originally native to Africa. It has very large, distinctive horns, these being even more impressive than those of the Texas Longhorn, for example, for besides being long, they are very thick in their circumference. Their digestive systems seem able to cope with poor quality and limited quantities of food and water. Their diet includes grasses, leaves and acorns. This great resilience has allowed them not only to survive for centuries in Africa but also to become established in Europe, South America, Australia and North America. Their life expectancy is more than 20 years.

The animals' large horns are honeycombed with blood vessels, and are used to thermo-regulate them in high temperatures. Blood moving through the horns is cooled by moving air, then flows back into the body lowering the body temperature. The Ankole-Watusi is medium in size, with cows weighing

The Ankole-Watusi's most striking feature is its impressively long horns.

900–1,200lbs (400–545kg) and bulls 1,000–1,600lbs (450–725kg). Newborn calves weigh 30–50lbs (14–23kg) and remain small for several months. This small birth-weight makes Ankole-Watusi bulls useful for breeding to virgin heifers of other breeds. During the day, the miniature calves sleep together, while another cow, not necessarily its mother, protects them. At night, the herd-members sleep together, with the calves in the centre of the group. The horns of the adults serve as formidable weapons against any potential intruders.

This ancient breed of long-horned, humpless domestic cattle were well-established in the Nile valley by 4,000 BC, when they were known as Egyptian or Hamitic Longhorns, appearing in pictographs in Egyptian pyramids. Over the next 2,000 years, they migrated from the Nile into Ethiopia, then down into the southern reaches of Africa. By 2,000 BC, the humped Longhorn Zebu, from Pakistan and India, had reached Africa, and on reaching Ethiopia and Somalia, interbred with

Ankole-Watusi were rarely used for meat, since an owner's wealth was counted in live animals.

the Egyptian Longhorns. This became the hybrid Sanga, which spread to the Sudan, Uganda, Kenya, and other parts of eastern Africa, becoming the base stock of many of the indigenous African breeds. The Sanga demonstrated most of the typical Zebu characteristics, such as the pendulous dewlap, upturned horns, and a neck hump of variable size. Modern descendants of the Sanga, however, vary greatly in size, conformation and horns, due to differing selection pressures exerted by different tribes.

Particularly remarkable are the cattle found in Uganda, Rwanda and Burundi. In Uganda, the Nkole tribe's Sanga variety is known as the Ankole, while in Rwanda and Burundi, the Tutsi tribe's Sanga is called the Watusi. The Rwanda common strain of Watusi is called the Inkuku, while the giant-horned strain, owned by the Tutsi kings and chiefs, is called the Inyambo, though some current tribal reports claim that this type is now extinct. Traditionally, Ankole-Watusi cattle were considered sacred. They supplied milk but were only rarely used for meat, since an owner's wealth was counted in live animals. Under traditional management, the Ankole cow was grazed all day, then brought home to her young calf. The calf was allowed to suckle briefly to stimulate milk flow, then the cow was milked by the herdsman. The calf then suckled after hand-milking was finished and was again separated from its mother. The process was repeated the next morning. This minimal nourishment of calves resulted in high death rates in the young. Moreover, milk production was not high, with a typical cow producing only two pints of milk daily, although an exceptional one could manage up to eight pints. In addition, the lactation period was short. Over the last ten years, the national government has attempted to select for animals which produce more milk and produce better meat, although famine and disease, as well as the conflict with traditional practices, have slowed this effort.

AUBRAC

An ancient French breed of beef cattle, Aubracs have long, lyre-shaped horns. They are tan in colour, the nose and eyes encircled with white. Cows weigh 1,320lbs (600kg) with bulls coming in at 1,980lbs (900kg). This is a tough breed, with high resistance to disease and great longevity; females calve easily.

The Aubrac has 150 years of breeding history. After centuries of selective breeding by French monks, it was decided to open a herd book in 1893, which began the golden age of the breed; cows spent the winter in the farm and gave birth to calves. At the end of spring the cows went

up into the mountains, where rich grasses and flowers produced the best milk to make Laguiole cheese, and young bulls were bred to became draft animals. There existed 300,000 Aubracs at this time, but their popularity was to decline by the end of the Second World War. By the 1950s, milk could be more cheaply produced by new breeds, such as Braunviehs or Holsteins, and with their previous niche now filled by these, the Aubrac became a beef breed.

Today, there are around 3,000 cows in the French stud book, the whole French population of Aubracs representing approximately 10,000. In Germany there are 2,500 cows, half of them included in a stud book, most of which are kept as sucklers. In France, 60 per cent are covered by Charolais bulls, because of the Aubrac's ability to cross successfully with heavy beef breeds. With the financial help of local authorities, however, small groups of farmers still choose to keep a few Aubracs for dairy use.

The Aubrac is distinguished by its outward-facing, lyre-shaped horns.

AYRSHIRE

A breed of dairy cattle originating in Ayrshire, Scotland. The average mature Ayrshire weighs 1,000–1,300lbs (450–600kg). It is a hardy breed, with a white coat splashed with large patches of bay (red-brown) to brown colouring. It has naturally long lyre horns which are usually removed when calves are young. They are known for their ability to convert grass into milk efficiently, and for their hardiness. The breed's strong points are the now-desired traits of easy calving and longevity. They also have 'spirited' natures, which may or may not be regarded as desirable.

Also known as Dunlop or Cunninghame cattle, they were once exported to all parts of the world, and extensive cattle docks once existed at Cunninghamhead station for loading and export purposes. The Dunlop family is credited with developing this line, with animals brought in from Holland.

These are strong, rugged cattle that adapt to all management

Ayrshires are known for their hardiness, longevity and easy calving.

systems, including group handling on dairy farms with free stalls and milking parlours. Ayrshires have excellent udder conformation and are not subject to excessive foot and leg problems, traits which make them outstanding commercial dairy cattle. Other traits that make them

ABOVE & OPPOSITE: Ayrshire cattle from Scotland and Sweden have been imported to Finland since 1845, and have been crossed with other breeds to produce the Finnish Ayrshire.

attractive to the commercial dairyman include the vigour of Ayrshire calves, which are strong and easy to raise. The Ayrshire is a moderate butterfat milk producer and relatively high-protein breed.

Ayrshires (especially those in Finland) are also crossbred with Holstein cattle in order to improve the Holstein's hardiness and fertility.

BEEFALO

A fertile hybrid offspring of the domestic cattle, *Bos taurus*, and the American bison, *Bison bison* (generally but erroneously called

buffalo in the US). It was created to combine the best characteristics of both animals with a view to beef production.

Accidental crosses were noticed as far back as 1749 in the southern English colonies of North America, while cattle and buffalo were first intentionally crossbred during the mid-1800s, with Charles Goodnight being one of the first to succeed. Charles Jesse 'Buffalo' Jones also worked to cross buffalo and cattle with the hope that they would survive the harsh winters. He called the result 'cattalo' in 1888. Mossom Boyd of Bobcaygeon, Ontario, first started the practice in Canada. After his death in 1914, the Canadian government continued experiments in crossbreeding up to 1964, with little success. Lawrence Boyd continues the crossbreeding work of his grandfather on a farm in Alberta.

It was found early on that crossing a male buffalo with a domestic cow would produce few offspring, but that crossing a domestic bull with a buffalo cow apparently solved the problem. The female offspring proved fertile, but the males were rarely so. Although the cattalo performed well, the mating problems meant that the breeder had to maintain a herd of wild and difficult-to-handle buffalo cows.

In 1965, Jim Burnett of Montana produced a hybrid bull that was fertile. Soon after, Cory Skowronek of California formed the World Beefalo Association and began marketing the hybrids as a new breed. The new name, Beefalo, was meant to separate this hybrid from the problems associated with the old cattalo hybrids. The breed was eventually set at being genetically at least 5/8 *Bos taurus* and 3/8 *Bison bison*, while animals with higher percentages of bison genetics are known as 'bison hybrids'.

A US Department of Agriculture study showed Beefalo meat, like bison, to be lower in fat and cholesterol. The association claims that Beefalo are better able to tolerate cold and need less assistance calving than cattle, while having domestic cattle's docile nature and fast growth rate; they are also thought to produce less damage to rangeland than cattle.

Creating the Beefalo, however, has proved to be a serious setback to wild American bison conservation. The American bison population has been growing rapidly, and is estimated at 350,000, but this is compared with an estimated 60 to 100 million in the mid-19th century. Most current herds, however, are genetically polluted or partly crossbred with cattle, and hence are in fact 'beefalo'; today there are only four genetically unmixed American bison herds left, and only one that is also free of brucellosis; it roams Wind Cave National Park.

BEEFMASTER

A breed of beef cattle developed in the early 1930s by Tom Lasater, from a crossing of Hereford and Shorthorn cattle with Brahman stock. The exact mixture of the foundation cattle is unknown, but is thought to be about 25per cent Hereford, 25 per cent Shorthorn and 50 per cent Brahman. It was the second new breed of cattle registered in the United States. The original intention was to produce cattle that could produce economically in the difficult environment of southern Texas. The cattle were selected by using the six essentials: weight, conformation, milking ability, fertility, hardiness and disposition. Though there are no standards for colour, most are red, although others are paint, dun, roan, white, brown, tan or black.

BELGIAN BLUE

A beef breed from Belgium, known there as La Race de la Moyenne et Haute Belgique. Alternative names include Belgian Blue-White, Belgian White-and-Blue Pied, Belgian White-Blue, and Blue Belgian. The sculpted, heavily muscled appearance is known as 'double muscling', and it is a trait shared by the Piedmontese breed. Belgian Blues are named for their typically blue-grey mottled hair-colour, although it can also be white or black. Critics refer to them as 'monster cows', and some governments, including that of

The 'double-muscled' Belgian Blue.

Denmark, have advocated that the strain be eliminated.

The breed originated in central and upper Belgium in the 19th century from crossing local cattle with Shorthorns from the United Kingdom and possibly Charolais from France. At first there were both milking and beef strains of the breed. The modern beef breed was developed in the 1950s by Professor Hanset in Liège.

OPPOSITE: Belgian Blue cattle are named for their typically blue-grey mottled hair colour, although they can also be white or black.

BELOW: A prize Belgian Blue bull in the show ring.

The Belgian Blue has a natural mutation of the gene that codes for myostatin, a protein that counteracts muscle growth. The truncated myostatin is unable to function in this capacity, resulting in accelerated lean muscle growth, due primarily to hyperplasia rather than hypertrophy. This defect is maintained through linebreeding. The mutation also interferes with fat deposition, resulting in very lean meat. Cows bred to double-muscled bulls are often unable to give birth naturally, requiring Caesarean sections instead. Double-muscled cows can also experience difficult births, even when bred to normal beef bulls or even dairy bulls, because of narrowing of the birth canal.

BLONDE D'AQUITAINE

The second most populous breed in France, this is a breed of beef cattle originating in the Aquitaine region of south-west France, and which embraces the area of the Garonne valley and the Pyrenees. The breed is a combination of three local strains, the Garonnais, the Quercy and the Blonde des Pyrénées. Blondes had always been hardy, lean animals, with

OPPOSITE & BELOW: The Blonde D'Aquitaine: its muscular appearance is a product of its early development as a draught animal.

a light but strong bone structure, but they were predominantly used as draught animals until the end of the Second World War, which resulted in muscle development, hardiness and docility. Blondes show some variation of colour ranging from almost white to tan.

BRAFORD

Used primarily for beef but occasionally in rodeo, due to its massive bulk and bone density, hardiness, heat endurance, and arguably bad-tempered disposition.

Brafords were developed in Florida through crossings between Hereford bulls and Brahman cows, and they carry the characteristics of both parents. The Braford is red, like a Hereford, with white underbelly, head and feet. It is stockier than a Hereford, however, having inherited this feature from the Brahman. A separate, unrelated bloodline was also formed in Australia.

Brafords have good heat- and insect-resistance because of the increased number of sweat glands and the oily skin coming from their Brahman heritage. They do as well in warm climates as they do in more northerly regions.

ABOVE: Brafords were originally bred in Florida, USA, but there is also a strain in Australia. The breed is particularly resilient to heat.

OPPOSITE: The Brahman is sacred in India.

BRAHMAN

These are the 'sacred cattle' of India, and many of the Hindu faith will neither eat their meat, nor will they permit them to be slaughtered. The Brahman is a breed of Zebu cattle (*Bos indicus*), that was eventually exported from India to the rest of the world. The main breed was the Kankrej, called Guzerat in Brazil. Also used were the Nelore or Ongole and the Gir or Gyr.

The American Brahman was the first breed of beef cattle developed in the United States. The American Brahman Breeders Association was formed in 1924 as the official herd registry to track and verify bloodlines, the name 'Brahman' having been created by the association's first secretary, J.W. Sartwelle. The American Brahman was the product of a nucleus of approximately 266 bulls and 22 females of several *Bos indicus* types, imported into the United States between 1854 and 1926. It has a distinctive large hump, and a loose flap of skin (dewlap) hanging from the neck. Ears are larger than *Bos taurus* breeds. Bulls weigh in at 1,600–2,200lbs (725–1000kg), with

cows coming in at 1,000–1,400lbs (450–635kg). American Brahmans are known to be a docile, intelligent breed. They can be grey or red in colour, the tail switch being black, and there is black pigmentation on the nose, tips of ears, and hooves. They are primarily horned cattle, although some bloodlines are naturally polled (without horns).

Brahmans can withstand heat better than European cattle. This is because their abundance of loose skin is thought to increase the body surface area exposed to cooling. They also sweat more freely, which contributes materially to their tolerance of heat. They are known for their longevity and resistance to parasites and disease. For these reasons, Brahmans have been extensively

ABOVE & OPPOSITE: Brahmans can withstand heat better than European cattle.

crossbred with European (*Bos taurus*) cattle in subtropical regions of the world. Brahman-crossed cattle, referred to as F-1 Brahmans, are most popularly crossings with Hereford and Angus cattle.

Brahman cows make extremely good, protective mothers. At birth, calves weigh 60–65lbs (27–30kg), but put on weight rapidly due to the outstanding quality of the milk. In some countries, especially South America, Brahman cattle are used for both milk and beef production.

The Brahman is one of the most popular breeds intended for meat processing and is widely used in Argentina, Brazil, the United States, Colombia and northern Australia (especially Queensland, the Kimberley (Western Australia) and the Northern Territory) among many other places. It has been used to develop numerous other US beef breeds, including the Brangus, Beefmaster, Simbrah and Santa Gertrudis.

BRAUNVIEH/BROWN SWISS

Braunviehs are the original 'brown cattle' of Switzerland, known for their docility, excellent milk production and the quality of their meat. Braunvieh cattle, imported into the United States in the 19th century, were the origin of the modern Brown Swiss cattle breed. Since the 1960s, however, Brown Swiss cattle have been crossed back into the Braunvieh stock of Europe. Original Swiss Braunvieh were also imported directly from Switzerland in 1983, which is the first time since 1880, by Harlan Doeschot of Firth, Nebraska, who had been in Switzerland looking for Simmental cattle to import and was greatly impressed by the uniformity and reproduction efficiency of the Braunvieh breed.

ABOVE & OPPOSITE: Braunvieh are the archetypal 'brown cattle' of Switzerland.

Coats come in various shades of brown, with lighter points.

BUELINGO

The Buelingo is a striking breed, its white belt dividing its black or cherry-red forequarters and hindquarters. The belt size varies but must be located between the hip bones and the shoulder blades and encompass the entire body without a break. Other colours, such as grullo- and dun-belted, are also eligible for registration.

A composite breed, the distinctive belt is largely derived from the Dutch Belt dairy cattle, with infusions of Scotch Highland, Belted Galloway, Angus, Chianina, Limousin and Shorthorn.The cattle can be polled or horned. The breed was developed on the Bueling Ranch in Ransom County, North Dakota, by Russell Bueling and Professor R.B. Danielson of the Animal Science Department of the North Dakota State University at Fargo.

The breed is well-accepted all over North America, and is now distributed throughout 30 states and three Canadian provinces. The founders of the breed followed a planned programme to develop a breed of cattle that suits today's needs, these being good meat quality of excellent taste with modest fat marbling. Breed traits are fertility, desirable calving weights, vigorous calves with rapid maturation, excellent maternal instincts and good dispositions. This is a moderately-sized breed, with males weighing in at 1,800–2,000lbs (800–900kg), with females coming in at 1,000–1,200lbs (450–540kg). The Buelingo breed is as well-adapted to the vast range areas of the American West as it is to the lush farms of the south and east. It is the perfect breed for those who must show a profit, but who may not necessarily have large acreages at their disposal.

Cows reach puberty at an early age, are extremely fertile, and are

known for their docile dispositions and for the ample supply of rich milk that they produce. Birth weights of calves are approximately 72lbs (33kg) with 205-day weaning weights running as high as 650–770lbs (295–350kg).

CANADIAN SPECKLE PARK

The Canadian Speckle Park is one of only a few beef cattle breeds developed in Canada, being native to the province of Saskatchewan. The latter half of the breed's name derives from the characteristic white, black and grey spots and patches of colour that appear on the coat.

This is one of the newest of the cattle breeds, having been officially recognized by the Canadian government in 2006. A breed association was formed in 1985, and exports of Canadian Speckle Parks to the US and Australia have now been made.

Work on developing the breed began in the 1950s, when a crossing of a crossbred heifer of roan

RIGHT & OPPOSITE: Braunviehs in Switzerland wear cow bells, so that they can't run away or wander off without being heard.

Shorthorn descent was made with a Black Angus bull. The spotting became a dominant trait in the offspring, and the beginnings of a new breed emerged. Descended solely from British beef breeds, the Canadian Speckle Park is naturally polled and inherits many of the characteristics of the Angus.

CANADIENNE

The Canadienne is a dairy breed developed in francophone Canada from cattle imported back in the 17th century from Brittany and Normandy. It is the only breed of dairy cow native to North America, being the result of genetic blends, and was once exclusively and intimately linked with Quebec's terroir. Today the Canadienne is seriously threatened due to cross-breeding with the Brown Swiss breed. Very few purebred animals remain, but a current trend towards the traditional-type Canadienne has prompted conservancy and development efforts with regards to its genetic make-up, and to guarantee a future for this French-Canadian heritage breed.

The founding of Quebéc in 1608 marked the beginning of a great adventure where cattle in Canada were concerned. It is likely, however, that most animals introduced from that date onwards were destroyed during the siege of Québec, between 1629 and 1632. Champlain's return in 1633 saw bovines permanently reintroduced to the land, but it appears that the more sizeable imports occurred between 1660 and 1670, under the Colbert and Talon administrations. In fact, this period is characterized by rapid cattle population growth in New France.

It seems that bovine imports ceased after this period, probably because they could be bred, and thereby their numbers multiplied. Due to harsh climate, traditional-type cattle would have faced supreme challenges upon arriving in Canada, and reproduction would have operated on a natural selection basis, which contributed to making the Canadienne a breed unique to the Americas and specific to Québec.

The Canadienne represented the majority of dairy cattle until 1850, when it began to suffer from the rivalry of other newly imported breeds. In 1886, when faced with the threat of extinction, breeders decided to create the Canadienne cattle breed's herd book, with the establishment of certain specific standards. In 1850, there were at least 300,000 Canadienne cows, but the breed's numbers fell into significant decline from this point on until, by 1970, there were only 5,000–10,000. In regards to a payment system for milk based on volume, the Canadienne breed was simply overcome, likely due to the rivalry of other breeds. In fact, it was during this decade that a crossbreeding programme with the Brown Swiss breed was pushed forward to improve conformity and production, also presumbly to allow the Canadienne to maintain and even increase its numbers. The Canadienne did improve, but the breed's numbers did not enjoy the success anticipated. In brief, it appears that crossbreeding results were disappointing. Today the breed's numbers fluctuate around 1,000 heads, but most have less than 93.75 per cent pure Canadienne blood (percentage required for purebred status), and possess Brown Swiss genes. Purebred or pureblooded numbers are fewer than 250 females, of which fewer than 100 have never been crossbred. The Canadienne cattle breed continues to lose ground and its purebred population is in a critical state, threatened with extinction. In Quebec, several individuals have recently become aware of the breed's historical and heritage relevance and value. In Charlevoix, where 35 per cent of all purebred females are to be found and the population is simply thriving, the purebred breed is at the heart of a development project involving its milk. Should the Canadienne breed be able to regain its lost importance is quite improbable, but the breed could nevertheless find a vocation that guarantees a more promising future for itself.

CHAROLAIS

Charolais is a breed named after the Monts du Charrolais hills in eastern France where it originated. Charolais are raised for their meat and are known for their composite qualities when crossed with other breeds, most notably Angus and Hereford. The breed tends to be large and well-muscled, with males weighing up to 2,500lbs (1130kg) and females coming in at 2,000lbs (910kg).

The Charolais was introduced into the southern USA as early as the 1940s, this being the most popular after the English breeds and the Brahman, with which it was often crossed. It was known to produce beef that had more lean meat and less fat. In the 1970s Charolais crossbred steers won at a number of prominent shows, particularly in Texas, the first being the San Antonio Livestock Show in 1971.

Charolais cattle originated in the hills of eastern France. Legend has it that white cattle were first noticed in the region as early as AD 878, and by the 16th and 17th centuries were well-known in French markets, especially at Lyon and Villefranche.

This breed has been quite popular in the north of Australia, where they are used for crossbreeding, and it has also gained popularity in the southern United States, where Charolais, often crossed with other breeds, have increasingly replaced Herefords. Despite their relatively northerly origin, Charolais tolerate heat well, and show good weight gains even on mediocre pasturage.

The coat is almost pure white, but Australian and Canadian breed standards also recognize cattle possessing a light red colour, when they are known as 'Red Factor' Charolais. Charolais may also be black in colour. The term Charbray refers to the offspring of Charolais crossed with Brahman and is recognized as a breed in its own right.

ABOVE & RIGHT: Charolais are usually pure white, although a light red colour and even black are also possible.

0252
0041

CHIANINA

Chianina is an Italian breed of beef cattle, being one of the oldest of its kind existing in the world. The famous *bistecca alla fiorentina* is produced from its meat. Originating in the Chiana Valley, the breed dates back to the time of the Roman Empire, the Chianina being originally developed as a dual-purpose breed for meat and draught purposes. The herd book dates back to 1856. A Chianina calf is the prize awarded to the winning team at the *Calcio Storico Fiorentino* football competition in Florence, which dates from the 15th century and which is held every year.

Chianina have white hair and a black switch. They have black skin pigmentation. They are heat-tolerant

and have a gentle disposition. In size, they are the largest breed, the males generally standing around 6ft (1.8m) tall and weighing up to 3,835lbs (1740kg), which was the weight of Donetto, when he was exhibited at the Arezzo show in 1955. Females stand at 5ft (1.5m) or so, weighing in at up to 2,400lbs (1090kg).

Chianina cattle are used in breeding programmes not only for their excellent growth-rate and high-quality meat, but also for their tolerance of heat. They are also great foragers. Moreover, they are also tolerant of diseases and insects to a much greater degree than many other domesticated cattle, making them popular in many countries beyond their native land.

The Chianina may well be one of the oldest breeds of cattle in existence. They were praised by the Latin Georgic poets, Columella and Vergil, and were the models for Roman sculptures.

CORRIENTE

The breed is descended from Spanish cattle brought to the Americas in the late 1400s. Today, they are primarily used in rodeo events, such as team roping and bulldogging (steer-wrestling). Some breeders raise them for their meat, which is significantly leaner than that of most modern beef cattle varieties .

The Corriente is a fairly small animal, with the female averaging well under 1,000lbs (450kg). Corrientes are lean, athletic, and have long, upcurving horns. They are known to be 'easy-keepers', in that they eat significantly less than bigger beef breeds, and require little human assistance when calving. Like Texas Longhorns (which many believe to be descended from Corrientes), they require less water and can live on sparse open range. Corrientes are also accomplished escape artists,

having the ability to clear a standard barbed-wire fence and squeeze through fairly small openings.

Names for the breed differ from place to place. The official breed registry in the United States uses the name Corriente, which is also most commonly used in northern Mexico. In other parts of Mexico, they are known as Criollo or Chinampo, and they are closely related to Pineywoods and Florida Cracker cattle, two breeds from the Gulf Coast and Florida.

DEVON AND MILKING DEVON

The Devon is a breed of cattle originating in England's south-west, primarily in the counties of Devon, Somerset, Cornwall, and Dorset. The Devon, together with the Hereford, Sussex, Lincoln Red and Red Poll, is one of several modern breeds derived from the ancient red cattle of southern England, and which existed in pre-Roman times. It is a rich red colour, which gave rise to other names, such as Devon Ruby or Red Ruby. The breed is also sometimes referred to as the North Devon to avoid confusion with the yellowish-brown South Devon breed.

In 1623, the ship, *Charity*, brought a consignment of red cattle (one bull and three heifers) from Devon to Edward Winslow, the agent for Plymouth Colony, which may have been of North Devon type.

OPPOSITE: The Devon is a rich red colour and is often called the Red Ruby or Devon Ruby.

ABOVE: A Milking Devon bull.

Although cattle had already been imported to the continent by the Spanish at an earlier date (descendants of which are the Texas Longhorn, Pineywoods and Florida Cracker breeds), this was the first arrival of British stock in the Americas, where early improvers of the breed were Francis Quartly and his brothers William and Henry, and John Tanner Davy and his brother William.

Colonel John Tanner Davy founded the Devon herd book in 1850, which the Devon Cattle Breeders' Society took over in 1884. Although the Devon was originally a horned breed, American stockmen developed a polled strain of purebred Devons that can be traced back to the bull Missouri 9097, a hornless individual born in 1915 in the Devon herd owned by Case and Elling in Concordia, Missouri.

The Devon was previously classified as a dual-purpose breed. Over the past half-century, however, it has developed as a beef-type breed. The rate of maturity has been accelerated, and the common criticism of light hindquarters and sickle hocks have been reduced to a minimum. Devons have become longer, taller and trimmer, but not to such extremes as in some other breeds. The traditional multi-purpose animals still exist in the US and are now known as Milking Devons, although they are very rare. They are registered with the American Milking Devon Cattle Association. Today, Milking Devons are still one of the most endangered breeds of cattle in the world.

Modern North Devons have been bred to be used almost exclusively for beef production, while Milking Devons are a multi-purpose animal akin to the stock which made that first transatlantic journey. Despite their name, they are also suited to meat production and to work as draught animals (i.e. oxen), being active, intelligent and relatively strong for their size. Considered to be one of the oldest and purest breeds of American cattle in existence, Milking Devons are also exceedingly rare.

The American Milking Devon is now one of only a few truly triple-purpose cattle breeds left in the West, being valued for meat, milk and draught. They are medium-sized animals, the males averaging 1,600lbs (726kg) and the females 1,100lbs (500kg). The coat is a dark, glossy red and the horns are white, ideally with black tips.

DEXTER

Dexters are the smallest of the European cattle breeds, being about half the size of a traditional Hereford and about a third that of a Friesian milking cow. Like the Kerry, they are descended from the predominantly black cattle of the early Celts. The Dexter breed originated in south-west Ireland, arriving in England in 1882, where it was maintained as a pure breed in a number of small herds, having by now virtually disappeared from Ireland.

The mature females weigh between 600–700lbs (270–318kg) and mature males about 1,000lbs (450kg). Despite their small size, the body is wide and deep with well-rounded hindquarters. Although black, dark-red or dun Dexters can be found, they are always of a solid colour, with only very minimal white marking on udders or behind navels. Horns are

Dexters are small cattle, descended from the predominantly black cattle of the early Celts.

rather small and thick and grow outwards with a forward curve on the male, and upwards on the cow.

Once very rare in both the UK and the US, Dexters have been enjoying something of a resurgence in both countries. These gentle, hardy and easy-to-handle animals require less pasture and feed than other breeds. They thrive in hot as well as cold climates and do well outdoors all-year-round, needing only a windbreak, shelter and fresh water. Fertility is high and calves are dropped in the field without difficulty. They are dual-purpose, being raised for both milk and meat. Sometimes, however, Dexters are listed as a triple-purpose breed, since they are also used as oxen. The success of the Dexter over the last

25–30 years is quite outstanding, its ability to adapt to varying and extreme climatic conditions and to different systems of management being typically characteristic.

They have established themselves well in many other parts of the world, having been exported to Australia, New Zealand, Cuba, Argentina, Kenya, Zimbabwe, Italy, Belgium, Denmark and Germany. Dexters produce a rich milk that is relatively high in butterfat (4 per cent) and the quality of the milk overall is similar to that of the Jersey. Some claim the milk is more naturally homogenized than others, due to the smaller fat globules in its make-up. Dexters can reasonably be expected to produce 2–2.5 gallons (7.6–9.5 litres) per day. Beef animals mature in 18 months, the result being small cuts of high-quality meat with little waste. The beef is well-marbled and a slightly darker red than that of other breeds.

The cows are exceptionally good mothers, being very protective of their calves. They produce enough milk to feed 2–3 calves, and will willingly nurse calves that are not their own. They are known for easy calving, and this trait, along with the small size of their calves, has produced a small but growing market in the US for Dexter bulls to breed to first-calf heifers, among the larger beef breeds, to eliminate problems at parturition.

Dexters make excellent mothers and will even suckle calves that are not their own.

DUTCH BELTED

The origin of Dutch Belted or Lakenvelder cattle is not precisely known, but they have existed as a pure breed since the 1600s, where they were found principally in the ownership of the Dutch nobility. According to records, this is the only belted breed that can be traced directly back to the original belted or 'canvassed' cattle described in Switzerland and Austria. These were called Gurtenvieh by the Dutch, and were evidently brought from the mountain farms of Appenzell and the Tyrol during or soon after the feudal period, reaching their heyday in the Netherlands in around 1750. The Dutch were very protective of their belted cattle, highly prizing them for their milking and fattening abilities. Nowadays, the cattle have become too rare to be a popular type of beef.

The name Lakenvelder or Lakenfield derives from the word *laken*, meaning a sheet or cloth, and refers to the band passing around the body. In some countries, such animals with this marking are known as 'sheeted' cattle. This belt or sheet is of pure white hair, extending from the shoulders to the hip bones, and must encircle the body completely. The cattle are otherwise solid black (or occasionally red). In their original

Dutch Belted cattle may be traced back to the 1600s when thay were owned by Dutch nobility.

(9000 litres) of milk per lactation. But they are still critically rare, with only a few hundred head in the United States. However, there are small-scale initiatives to preserve the race, such as that of Minnesota Zoo, which is working to maintain the breed's survival.

There is a rare breed of domestic poultry, also called Lakenvelder, that has a solid black neck hackle and black tail but with a pure white body.

form they were horned and primarily a dairy breed, comparing favourably with the Holstein as far as milk yield was concerned.

By the 1930s Lakenvelder numbers were very low in the Netherlands and the herd book was closed. In the 1970s, however, a trust was set up to save the breed from extinction in its own country, even though it had been exported to the US as early as 1838, where it is still popular. Current races are more productive, with some Dutch Belted cows producing over 2,378 gallons

FLORIDA CRACKER

A breed of cattle native to the state of Florida, and named for the Florida Crackers who kept the breed. The Florida Crackers were the pioneer settlers of the state of Florida, who arrived in 1763 when Spain traded Florida to Great Britain. Other related breeds include the Corriente and Texas Longhorn.

The Florida Cracker is one of the oldest breeds of cattle in the United States, being the descendant of Spanish cattle brought by the conquistadors to the New World, beginning in the early 1500s. As the Spanish colonized Florida and other parts of the Americas, they established the low-input, extensive cattle-ranging systems typical of Spanish ranching. The Florida Cracker, and others which developed under these conditions, are known as Criollos, which are South or Central American domestic breeds of Spanish origin.

The breed was shaped primarily by natural selection in an environment that is generally hostile to cattle. This resulted in a breed that is heat-tolerant, long-lived, resistant to parasites and diseases, and

productive on the low-quality forage found on the grasslands and in the swamps of the Deep South. It was not until the importation of Zebu from India and the development of the American Brahman breed in the 1900s that the Florida Cracker had competition from other heat-tolerant cattle. Not long afterwards, the development of parasiticides and other medications allowed British and European breeds to survive in the area, and the Florida cattle industry was further diversified.

This influx of new breeds very nearly caused the extinction of the Florida Cracker breed. By the mid-1900s, the majority of purebred cows had been crossbred, first to Brahmans and then to British and European breeds. Florida Crackers are small animals, with males weighing 800–1,200lbs (360–540kg) and females coming in at 600–800lbs (270–360kg). Horn-shapes vary, ranging from rather long and twisted to smaller and more crumpled. The colours of the cattle also vary from browns to whites to blacks, brindles, spotteds and roans.

Given that the breed is considered to be a living part of the state's history, Florida has been a leader in the conservation and promotion of the breed over the past few decades, but even so it is still quite rare and increased knowledge of the breed's existence is vital to its preservation. The state has supported the establishment of the Florida Cracker Cattle Association and a breed registry, which is operated by

The Florida Cracker is one of America's oldest breeds, being a descendant of cattle brought over by the Spanish conquistadors.

the American Livestock Breeds Conservancy. The Florida Cracker cattle breed is still quite rare, but its prospects are brighter than they have been for a long time.

GALLOWAY

The Galloway is one of the world's longest-established breeds of beef cattle, named after the Galloway region of Scotland, where it originated, but now found in many parts of the world. The Galloway was introduced into Canada in 1853, first registered in 1872, and the first registry of the breed was introduced in the USA in 1882.

The Galloway comes from the cattle native to an entire region of Scotland, so originally there was much variation within the breed, including many different colours. The original Galloway herd book only registered black cattle, but the recessive gene for red colour persisted in the population, and eventually dun Galloways were permitted. As a result, although black is still the most common colour for Galloways, they can also be red and several shades of dun.

The Galloway is naturally hornless, having a bone knob at the top of its skull that is called a poll. The breed's shaggy coat has both a thick, woolly undercoat for warmth and stiffer guard hairs that help shed water, making it well-adapted to harsher climates. Males weigh in at about 1,765lbs (800kg), while females come in at an average of 1,210lbs (550kg).

The Galloway is faced with the challenge of a genetic defect called tibial hemimelia (TH), caused by an abnormal recessive gene, and first

BELOW: The Galloway is naturally hornless, having a bone knob on the top of its skull instead.

OPPOSITE ABOVE: The Belted Galloway is affectionately known as a 'Beltie'.

OPPOSITE BELOW: The red colour in Galloway cattle is caused by a persistent recessive gene, so that although black is still the most common colour, Galloways can also be red or several shades of dun.

identified in Shorthorn cattle in 1999. TH is characterized by severe and lethal deformities in newborn calves, and those so affected are born with twisted rear legs with fused joints, and have large abdominal hernias and/or a skull deformity; they cannot stand to suckle and therfore must be destroyed.

In more recent times, two sister breeds to the Galloway have been created: the Belted Galloway and the White Galloway, both of which can be differentiated by distinctive colour patterns, and both can be either black, red, or dun.

The Belted Galloway features a wide, white stripe around the midriff, and is often affectionately referred to as a 'Beltie'. It was created by crossing Galloways with Dutch Belted cattle, a dairy breed. Belted Galloways are often smaller than regular Galloways, and often have more of a dairy or aesthetic focus.

The White Galloway is mostly white, with colour restricted to the ears, feet and around the eyes, and there will often be colour on the poll, tail and udder. The genes for this colour pattern were introduced by sources unknown.

BELOW: The White Galloway has colour restricted to the ears, feet and around the eyes.

OPPOSITE: The Gelbvieh is a native of Germany and is a honey-gold to red colour.

GELBVIEH

Gelbvieh ('Yellow cattle') is a dual-purpose breed originating in Bavaria, Germany, around the end of the 18th century. It is known both as the German Yellow and Einfarbiges gelbes Hohenvieh. In spite of the name, the breed is a honey gold to red colour, but due to crossbreeding with Angus cattle, a good portion are now black. The breed is usually horned. The Gelbvieh was partly derived from the Schwyz and Bernese breeds of Swiss cattle, having been based on quality of meat, milk, and strength as a draught animal.

The breed has been introduced to Australia, Canada, South Africa and the United States, primarily by artificial insemination.

Polled cattle have been developed in the United States from the use of naturally hornless foundation females. Proponents of the breed claim it has superior fertility, ease of calving, mothering ability, and growth-rate in the calves.

GUERNSEY

The Guernsey is a breed of cattle used in dairy farming. It is golden brown and white in colour, and is particularly renowned for the richness of its milk, as well as its hardiness and docile disposition. As its name implies, the Guernsey was bred on the British Channel Island of Guernsey. It is believed to be descended from two breeds brought over from nearby France: Isigny cattle from Normandy and the Froment du Léon from Brittany. The Guernsey was first recorded as a separate breed in around 1700. In 1789, imports of foreign cattle into Guernsey were forbidden by law to maintain the purity of the breed, although some cattle evacuated from Alderney during the Second World World were incorporated into the breed.

For a while, exports of cattle and semen were an important economic

resource, and in the early 20th century a large number of Guernsey cattle were exported to the United States. Today the breed is well-established in Great Britain, the United States, Canada, South Africa and elsewhere.

The unique qualities of the milk produced by the Guernsey cow have made the breed world-famous. The milk has a golden colour, due to an exceptionally high beta carotene content, this being a source of Vitamin A, which has been claimed to help reduce the risks of certain cancers. The milk also has a high butterfat content of 5 per cent and a high protein content of 3.7 per cent. Guernsey cows produce around 1,585 gallons (6000 litres) per cow per annum. In the US, Guernsey cows average 16,200lbs of milk per year with 4.5 per cent fat and 3.2 per cent protein. Guernsey cattle are known to produce the highest percentage of A2 milk of all breeds of dairy cattle.

From the 1950s to the early 1970s, Golden Guernsey trademark milk was sold in the US and Canada as a premium product, the golden colour being the most important marketing point and the source of the brand name. Only milk from Guernsey cows could be marketed under the Golden Guernsey trademark. However, the advent of homogenization and various changes to the way milk was priced and marketed, spelled the end of Golden

The Guernsey's milk is world-famous for its richness and high-protein content.

Guernsey-branded milk, although the trademark is still maintained by the American Guernsey Association today, and is used by various small-scale dairies around the country.

The Guernsey cow weighs in at 1,100lbs (500kg), slightly more than the Jersey cow, which is around 1,000lbs (450kg). Bulls weigh in at 1,300–1,545lbs (600–700kg), which is small by domestic cattle standards, and they can be surprisingly aggressive. The Guernsey cow has many notable advantages for the dairy farmer over other breeds. These include high efficiency of milk production, low incidence of calving difficulty and longevity. Inbreeding, however, is becoming a real concern due to the small gene pool in a given area, and may be solved in most cases by exchanging cows with no overlap in lineage from other farms. Guernsey cows are also sometimes regarded as somewhat more fragile than comparably-sized breeds.

HAYS CONVERTER

The first beef breed, recognized and registered as a purebreed under the provisions of the Canada Livestock Pedigree Act and developed by a Canadian livestock producer. Since the concept was first developed by the agriculturalist and politician Senator Harry Hays of Calgary, the objective was to produce superior animals with the selection based only on performance. Senator Hays had been considering the idea of a new beef breed for a long time; not only did he wish to develop a leaner breed, but he also wanted one that gained weight as efficiently as possible. The market demanded a steer in the range of 1,100lbs (500kg), therefore his goal, for maximum economic benefit, was to breed a beef animal that would reach this desirable market weight during its first year of life, when it could most efficiently convert feed to meat.

Senator Hays began his evolutionary quest in 1959 by carefully combining progeny from three outstanding animals, each from a different existing breed. These were: Fond Hope, a Holstein bull weighing 3,120lbs (1415kg), whose progeny were distinguished by their large size, rugged constitution and excellent, strong feet; Silver Prince 7P, a 2,640-lb (1200-kg) Hereford bull and a certified meat sire, noted for his ability to transmit size, length, bone and fleshing ability to his offspring; and Jane of Vernon, a 1,600-lb (725-kg) Brown Swiss cow famous for having what was judged to be the world's most perfect udder, among other virtues. Breeding continued between these three into several generations until Hays adjudged that the genetics were now right, and the herd was closed to all other outside breeding influence. By 1969, his own breed of cows had been bred to his own breed of bulls regularly and exclusively for seven years, and his work on improving nature's genetics was producing his anticipated results.

HEREFORD

The origin of the Hereford is lost in the mists of time but is generally agreed to have been founded on the draught ox descended from the small red cattle of Roman Britain and from a large Welsh breed once in evidence along the border of England and Wales. The Hereford Herd Book Society was founded in 1878 by J.H. Arkwright of Hampton Court,

OPPOSITE: The Hereford is a hardy ancient breed dating back to Roman times.

Herefordshire, under the patronage of Queen Victoria. The herd book has been closed since 1886 to any animal whose sire or dam has not been entered previously. So for over 100 years, the purity of the breed has remained intact.

Because of its performance as a crossing sire on commercial cattle and indigenous breeds in many countries, the impact of the Hereford on world beef production has been profound. This widespread popularity could only have come about because farmers, ranchers and feeders found the Hereford to be consistently profitable under a wide range of climates and conditions.

ABOVE, OPPOSITE & PAGE 68: Herefords are docile, long-living and easy to keep.

CATTLE

Fir0002/Flagstaffotos/Wikimedia Commons

More than five million pedigree Herefords exist in over 50 countries. The export of Herefords began in 1817, spreading across the United States and Canada, then through Mexico to the great beef-raising countries of South America. Today, Herefords dominate the world scene from the prairies to the pampas and from the Russian steppes to the South African veldt.

The Polled Hereford is a hornless but direct relative of the Hereford, being a natural genetic mutation that was selected into a separate breed. Breeding horned and hornless together functions as a genetic dehorner, and is often used as an alternative to the dehorning process, which causes stress and often weight loss

The first Herefords were exported to North America in 1817 and quickly became dominant over the traditional Longhorn and Shorthorn breeds. In 1898 Warren Gammon, an Iowa cattle rancher, decided to capitalize on the idea of breeding Polled Herefords. He located 1,500 naturally hornless cattle, then chose three bulls and eight cows from which he bred the strain known today as the

American Polled Hereford. Two organizations exist for Hereford breeders in the United States: the American Hereford Association and the American Polled Hereford Association.

The Hereford and its crosses are docile and have early maturation and longevity. They are low-maintenance, and have a good capacity to deal with grass and arable by-products, also the unique capacity to winter rough while holding their flesh. They put on weight faster than any other breed on pasture. No breed can equal the Hereford for siring profitable feeders, from dairy or beef cows of indigenous breeds anywhere in the world. The average weight of a mature male is around 2,530lbs (1145kg), while that of the mature female is around 1,890lbs (855kg).

HIGHLAND

The Highland cattle, or *kyloe,* has a long and distinguished history, not only in its native Scottish Highlands and Western Isles, but also in many far-flung parts of the world. One of Britain's oldest, most distinctive and best-loved breeds, it has remained largely unchanged for centuries, distinguished by its long, thick, flowing coat, coloured black, brindled, red, yellow or dun, and majestic sweeping horns. Written records go back to the 18th century and the Highland Cattle Herd Book, first published in 1885, lists pedigrees since that time.

Highland cattle are exceptionally hardy, and are able to convert poor

LEFT & PAGES 70–71: The Highland is a native breed from the Highlands and Western Isles of Scotland. It is very hardy and accustomed to rugged terrain.

grazing efficiently, eating plants that other breeds would avoid. They thrive where no other breed could exist, surviving on vast tracts of poor mountainous land, lashed by bitter winds and a high annual rainfall. They are remarkable for their longevity: many Highland cows continue to breed to ages in excess of 18 years, having borne 15 calves.

Highlands get most of their insulation from their thick, shaggy hair rather than from subcutaneous fat. Therefore they produce lean, well-marbled beef that is tender and succulent with a very distinctive flavour. It is healthy and nutritious, with lower levels of fat and cholesterol and a higher protein and iron content than other beef.

HOLSTEIN

The major historical development of this breed occurred in what is now the Netherlands and, more specifically, in the two northern provinces of North Holland and Friesland, which lay on either side of the Zuider Zee. The original stock were the cattle of the Batavians and Friesians, migrant European tribes which settled in the Rhine delta about 2,000 years ago. The Dutch breeders supervised the development of the breed, with the aim of obtaining animals which would make best use of grass, the area's most abundant resource. The result, over centuries of artificial selection, was an efficient, high-producing black-and-white dairy cow.

Holsteins are immediately recognizable by their distinctive colour markings and outstanding milk production. Holsteins are large animals, with colour patterns of black-and-white or red-and-white. A healthy Holstein calf weighs 90lbs (41kg) or more at birth, while a mature female weighs in at about 1,500lbs (680kg) and stands 58in (1.5m) tall at the shoulder.

Holsteins are good-natured, easy-to-handle and can be stabled without any problems. They are also resistant to stress, and are sociable within the herd. If managed well, they present no fertility problems, and they produce vigorous calves distinguished by rapid growth, early maturity and ease of rearing.

Holsteins are more than just a dairy breed. The animals also contribute to the meat supply worldwide, have a high growth per

CATTLE

centage in the fattening sector and produce meat with a fine fibre. In industries aimed exclusively at milk production, they are crossbred with beef breeds for a better quality veal. Holstein heifers can be bred at 15 months when they weigh about 800lbs (360kg), but it is desirable to have Holstein females calve for the first time between 24 and 27 months. Holstein gestation is approximately nine months. While some cows may live considerably longer, the normal productive life of a Holstein is six years.

With the settlement of the New World, markets began to develop for milk in America, and dairy breeders turned to Holland for their cattle. Winthrop Chenery, a Massachusetts breeder, purchased a cow, which had provided milk for the voyage, from a Dutch sailing master who had landed cargo at Boston in 1852. Chenery was so pleased with her milk production that he imported more Holsteins in 1857, 1859 and 1861, and many other breeders soon joined the race to establish Holsteins in America. After about 8,800 Holsteins had been imported, a cattle disease broke out in Europe and importation ceased. In the late 1800s, there was enough interest among Holstein breeders to form associations to record pedigrees and maintain herd books. These associations merged in 1885 to found the Holstein-Friesian Association of America. In 1994, the name was changed to Holstein Association USA, Inc.

JERSEY

Jerseys are a small breed of dairy cattle, thought to have been purebred for six centuries and originally bred on the British Channel Island of Jersey. Before 1789, cows were occasionally given as dowries in inter-island marriages between Jersey and Guernsey inhabitants. In that year, imports of foreign cattle into Jersey were forbidden by law to maintain the purity of the breed, although exports of cattle and semen have been an important economic resource for the island. This restriction had initially been introduced to prevent a collapse in the export price. The

Holstein cattle are easily recognizable by their distinctive colour markings. Their milk production is outstanding.

United Kingdom levied no import duty on cattle imported from Jersey, but cattle were being shipped from France to Jersey and then shipped onward to England to circumvent the tariff on French cattle. The increase in the supply of cattle, sometimes of inferior quality, was bringing the price down and damaging the reputation of Jersey cattle. The import ban stabilized the price and enabled a more scientifically- controlled programme of breeding to be undertaken. In July 2008, the States of Jersey took the historic step of ending the ban on imports, and allowing the import of bull semen from any breed of cattle, although only semen that is genetically pure will enable the resultant progeny to be entered in the Jersey Herd Book.

The Royal Jersey Agricultural and Horticultural Society was established in 1833, at a time when the breed displayed greater variation than it does today, with white, dark-brown and mulberry beasts. But

LEFT & OPPOSITE: The Jersey is an attractive small cow with a good milk yield that makes it popular in many parts of the world.

since the honey-brown cows sold best, the breed was developed accordingly. In 1860 1,138 cows were exported via England, the average price being £16 per head. By 1910 over 1,000 head were being exported annually to the United States alone, which now has the fastest-growing dairy breed in the world.

The breed is able to adapt to a wide range of climatic and geographical conditions, and Jersey herds are to be found in most parts of the world. They are excellent grazers and perform well in intensive grazing programmes. Each lactation, most Jerseys are capable of producing more than 13 times their body weight in milk. The milk has a high butterfat content (6 per cent butterfat and 4 per cent protein), and the lower maintenance costs due to lower body weight, as well as its amenable disposition, contribute to the Jersey's popularity. It also has high fertility, superior grazing ability, it calves easily and with a relatively lower rate of dystocia, which has led to the Jersey's crossbreeding with other dairy and even beef breeds to reduce calving-related injuries.

The cow's weight ranges from 800–1,000lbs (360–450kg), while bulls are also small, with weights ranging from 1,200–1,800lbs (540–820kg). They are notoriously aggressive. Jerseys come in all shades of brown, from light tan to almost black. All purebred Jerseys have a lighter band around their muzzles, a dark switch (long hair on the end of the tail), and black hooves, although in recent years regulations governing colour have been relaxed to allow a broadening of the gene pool.

Cows are calm and docile animals, but tend to be rather more nervous than other dairy breeds. They are also highly recommended cows for first time owners. Unfortunately, they are also notorious for a tendency towards parturient hypocalcaemia (milk fever) and for producing frail calves that require more attentive management in cold weather, due to their smaller body mass.

Castrated males can be trained into fine oxen which, due to their small size and gentler nature, make them popular with young teamsters. Jersey oxen are not as strong as larger breeds, however, and are generally out of favour with competitive teamsters.

KERRY

A rare breed of dairy cattle native to Ireland. The Kerry is probably descended from the Celtic Shorthorn, brought to Ireland from Europe as early as 2000 BC. They cause less damage to pastures in high rainfall areas than larger breeds, their agility enabling them to travel safely over rough ground. They were developed as a milking breed suited to the small subsistence farms of southern and western Ireland. By 1983 there were only around 200 pedigree Kerry cattle in the world, but numbers have since increased.

Kerry cattle were frst imported to the United States in 1818 and prospered in the 19th century,

BELOW: The Kerry is a very rare Irish breed.

OPPOSITE: The Limousin, an ancient breed of French cattle.

becoming scarce by the 1930s. Today there are only a few herds in North America, these being mostly recent imports into Canada.

Believed to be one of the oldest breeds in Europe, the Kerry's skull is very similar in formation to those of the ancient aurochs of the Stone Age (*Bos primogenus*), despite being smaller in size. The Kerry's coat is almost entirely black, with a little white on the udder, the horns being whitish with dark tips. Cows weigh in at about 770–880lbs (350–400kg) and produce 6,610–8,155lbs (3000–3700kg) of milk per lactation.

LIMOUSIN

Limousins are beef cattle originally bred in the Limousin and Marche regions of France. The breed may be as old as the history of Europe itself, for cave paintings of cattle, painted on the walls of the Lascaux Cave, in the Dordogne, France, bear a striking

resemblance to today's Limousin cattle. The breed originated in a rainy region with harsh climatic conditions and poor granite soil. Consequently, the growing of field crops was very difficult, and emphasis was placed on raising animals instead. As a result of their environment the cattle evolved into a breed of unusual sturdiness, health and adaptability. This lack of natural resources also enabled the region to remain relatively isolated, leaving the farmers free to develop their cattle with little outside

genetic interference. Limousins also gained a well-earned reputation as work animals in the early days of animal power.

The first herd book was established in 1886. Limousins are usually golden-brown, but other colours, such as black, have been developed through crossings with other breeds of cattle, in addition to which other traits, such as an absence of horns, have also been introduced. The Limousin is large, fine and has a strong-boned frame. Mature males weigh in at about 2,200lbs (1000kg), with mature females coming in at an average of 1,430lbs (650kg). The head is small and short with a broad forehead, and the neck is short with a broad muzzle. Limousins are known for their muscular build, feed-efficiency, ease of management and comparable calving ease. They also produce the leaner beef that is a requirement of the modern market.

Limousins are large and most often golden-brown in colour.

LINCOLN RED

The genetic base of the Lincoln Red is thought to be *Bos urus*, introduced into Britain by Viking invaders between AD 449 and 660. Little more was heard of these Lincolnshire cattle until 1695, when Gervaise Markham remarked on their quality in his book, *A Way to Get Wealth*. In the late 18th and 19th centuries, bulls selected from Durham and York shorthorns were crossed with these native cattle and the resulting progeny became known as the Lincoln Red. In 1799, the British Board of Agriculture described the Lincoln Red as 'a breed of cattle which is unsurpassed in this country for their disposition at any age to finish rapidly'. The breed was recognized in the first Coates Herd Book in 1822, and the Lincoln Red Association was subsequently formed in 1896. Lincoln Red cattle are dark red in colour and may be polled or horned.

In a move highly unusual in beef-breeding circles, the members of the North American Lincoln Red Association have banded together to preserve the genetic integrity of the breed. Consequently, the its gene pool value has been enhanced by the creation of several distinct new lines, in what will be an ongoing process.

LINEBACK

A dual-purpose breed of cattle, derived from the Holstein breed, to which the Randall cattle bear a strong resemblance. While not a commonplace breed, they are found in small numbers as 'productive pets' amid larger numbers of other cattle breeds on dairy farms throughout North America.

Known for their distinctive marking, the Lineback has a black base and nose, with a skunk-like white stripe or finching running along the spine. Due to the rarity of the pure-strain breed, variations, such as red linebacks and animals with heavily speckled pelts instead of the distinct stripe, occur from time to time.

Due to a propensity for producing large calves, that can result in dystocia, the breed is often crossbred to other dairy cattle, with the coloration trait being generally dominant in the offspring.

LOWLINE

Lowlines may appear to be dwarfs but this is not the case, in that the cattle have been specifically selected for this genetic trait. They are approximately 3.5ft (1m) tall and weigh up 1,300lbs (600kg). They are black in colour and naturally polled. They are claimed to be very docile, and are noted for their easy calving, with calves weighing 55lbs (25kg).

Lowline Angus cattle were developed from 1974 by the Trangie Agricultural Research Centre in

OPPOSITE: The Lincoln Red in thought to be the ancient *Bos urus*, brought to Britain by Viking invaders.

ABOVE RIGHT: The Lineback is named for the white stripe running down its back.

Australia. (The Angus breed has its origins in eastern Scotland, in the counties of Aberdeen and Angus, where it was developed from the native black hornless cattle.) The research was on the effects of genetic selection for growth rate to yearling age. Groups of Angus cattle, with high- and low-growth rates, were bred separately and then compared with a randomly selected control group. These groups were known as 'High

Line', 'Low Line' and 'Control Line'. The original stock was imported from Canada in 1929, with some Scottish and US animals added in subsequent years. When the experiment ended in 1992 the Lowline herd was sold to private breeders who formed the Australian Lowline Cattle Association. Lowline cattle are now bred in Canada, Australia, New Zealand, USA and China.

Lowline cattle are growing in popularity throughout the United States, used to produce beef or as docile pets. They are suitable animals for children to train and show.

MAINE-ANJOU

The foundation of the Maine-Anjou breed occurred in 1839, when the

French Mancelle was crossed with the English Durham, the purpose being to ensure better beef production. The Mancelle provided hardiness, vigour and excellent milking ability, even under sparse feeding conditions, while the Durham provided quality meat and rapid growth. In 1908 the cross was named Maine-Anjou and a breeders' society was formed, the name having been derived from the area in western France where the new breed was now flourishing. Cattlemen from Canada and the United States, upon seeing the breed in France, were impressed by its combination of growth, milking ability and docility, and this lead to the first Maine-Anjou importations into North America. They were first imported live into Canada in 1969, then later introduced into the United States through artificial insemination. The Maine-Anjou has evolved as a dual-purpose breed, with the cows used for milk production and the bull calves for meat.

Maine-Anjou cattle are red-and-white (sometimes black or roan) and are horned. They are a large breed, with males weighing in at 2,200–3,100lbs (998–1406kg), and females at 1,500–1,900lbs (680–862kg).

MARCHIGIANA

The Marchigiana is a breed of cattle, which some say was brought to Italy by barbarians after the fall of Rome in the fifth century. Native specifically to the Marche region, the Marchigiana is a large breed which today is kept for beef. Prior to the 1950s, however, it was also bred to perform draught work as an ox.

The Marchigiana was developed in the late 19th and early 20th centuries by crossing native Podolica cattle with the Chianina and Romagnola breeds. Today, the Marchigiana still bears a close conformational resemblance to the Chianina, although it is not as tall. Marchigiana cattle make up 45 per

OPPOSITE: The Maine-Anjou is the result of a crossing between a French Mancelle and an English Durham in 1839.

RIGHT: The Marchigiana is believed to have been brought to Italy by barbarians after the fall of Rome.

cent of the beef herd in Italy and they have been exported internationally, including to the United States and elsewhere. They are fast-maturing, come both polled and horned, and have short, light-grey to white coats. Occasionally they exhibit double-muscling, due to mutations in the myostatin gene.

MURRAY GREY

A polled breed of Australian beef cattle originating in the upper Murray River valley on the New South Wales/Victoria border. The Murray Grey occurred from an initial chance mating of a black Aberdeen Angus bull with a roan Shorthorn cow in 1905. The resulting 13 dun-grey calves were at first kept as curiosities and then bred on the Thologolong property along the

BELOW: The Marchigiana has a light-grey to white coat and a black nose.

OPPOSITE: The Murray Grey has dark-grey skin and a dun-grey coat.

Murray River by Peter and Ena Sutherland.

These unusually-coloured cattle grew quickly, were good converters of feed, and produced quality meat. Local cattlemen soon became interested in them and began breeding them. The first larger-scale commercial herds were established in the 1940s, while in the 1960s several cattle breeders were selling the cattle as commercial enterprises. The Murray Grey Beef Cattle Society was formed to register and administer the breed, performance-recording the herd using the internationally recognized Breedplan for monitoring growth, milk and carcase quality. There are also registries of the breed in Canada, New Zealand, the United Kingdom and the United States.

In 1963 negotiations were made to have the similar Tasmanian Grey beef cattle accepted into the Murray Grey Beef Cattle Society, but it was not until 1981 that the two organizations would be combined.

In the 1970s Greyman cattle were developed in Queensland specifically to suit the local environment, by combining Murray Grey and Brahman breeds. Breeders in the northern and western regions of Australia are increasingly using Murray Grey genetics to cross with *Bos indicus* cattle to improve fertility, docility and meat quality. Murray Greys are the third largest breed in Australia, and because of the beef's superior quality there is strong demand from Asian countries.

NORMANDE

Coming from the French province of Normandy, with a coastline on the English Channel, the breed is claimed to be descended from cattle brought to the area by Viking invaders in the 9th and 10th centuries. These cattle have subsequently evolved over 1,000 years into a dual-purpose breed to meet the milk, cheese and meat requirements of the residents of north-western France.

The present herd book was begun in 1883, and although the breed was decimated by the Allied invasion of Normandy during the

Second World War, there are currently three million Normande cattle in France.

Normandes are chestnut-brown pied or black pied. The head is white and the eyes are darkly 'spectacled' (described as having *lunettes* in France). Males are typically 2,425lbs (1100kg) in weight, while the females come in at 1,545lbs (700kg).

NORWEGIAN RED

The *Norsk rødt fe* was developed in Norway, the name often shortened to NRF. It has a red-and-white coat, and is primarily a dairy cattle breed.

The NRF, as a high-producing breed, was selected for a broad breeding objective, with increasing emphasis on functional traits such as health and fertility. The cattle may be either polled or horned and are noted for their hardiness and the richness of their milk. The breed was developed in the 1960s through crosses of dairy cattle with several Scandinavian breeds, including the Norwegian Red-and-White, the Red Trondheim and the Red Polled Østland. Later in 1963 the Døle was also absorbed into the designation and in 1968 South and West Norwegians were added. Other breeds which are said to have contributed to the gene pool include Ayshire, Swedish Red-and-White, Friesian and Holstein.By the mid 1970s the Norwegian Red was the dominant breed in its native country, and comprised 98 per cent of the cattle population.

Males weigh around 1,980lbs (900kg), with females coming in at about 1,090lbs (495kg); milk yields average 12,800lbs (5806kg) per lactation. Semen is frequently exported to the US dairy industry for crossbreeding with Holstein cattle.

OPPOSITE ABOVE & BELOW: Normande cattle.

BELOW LEFT & PAGE 88: The Norwegian Red was developed through crossings with several Scandinavian breeds in the 1960s.

PARTHENAIS

A beef cattle from the Deux-Sèvres department of western France. The name comes from Parthenay, a town that was an important cattle market during the Middle Ages. The golden age of this breed was the second part of the 19th century; a herd book was established in 1893, making it one of the oldest in France. At this time, all the Cognac vineyards had been destroyed by phylloxera, a pest louse affecting vines, and farmers brought in cattle in the meantime until the disease could be controlled. The famous Charente-Poitou butter, produced by the Parthenais, has the designation *Appellation d'Origine Contrôlée* (AOC). The breed has been exported to the United Kingdom, Ireland, the USA and Canada.

The Parthenais is golden brown in colour, with lighter eyes, muzzle and legs, while the nose, hooves and tail are black. Horns are crescent-shaped. Males weigh around 2,600lbs (1180kg), while mature females come in at around 1,600lbs (725kg).

Although the breed has a multi-purpose history, it has been selectively bred for beef since 1970, in that it has a degree of double-muscling, which produces healthy, good-tasting, lean meat.

PIEDMONTESE

A breed of cattle, said to be the descendants of aurochs (*Bos taurus*) and Zebu (*Bos indicus*), which migrated to the Piedmont region, in north-west Italy, some 25,000 years ago. The breed developed through natural selection followed by the normal processes of domestication and selective breeding. The cattle are raised both for their milk, which is used for a number of the region's traditional cheeses (Castelmagno,

RIGHT: The Parthenais is a golden brown in colour, with lighter areas around the eyes, muzzle and legs.

LEFT & OPPOSITE: Piedmontese are all horned in Italy, but in America may also be polled.

Bra, Raschera and Toma Piemontese), and for meat, which is regarded as a premium product.

All Italian white breeds, Piedmontese included, are born fawn or tan and change to the grey-white colour, with black skin pigmentation. The Piedmontese breed carries the gene for inactive myostatin, which not only increases muscularity but also reduces the fat content while improving tenderness in the beef. This low-fat beef is also lower in calories, higher in protein, and contains a higher percentage of the good Omega-3 fatty acid. The fullblood population is considered homozygous for this inactive myostatin gene. The beef from Piedmontese and Piedmontese-cross cattle is also consistent for these qualities of leanness and tenderness because it is a genetic influence rather than an environmental effect.

PINEYWOODS

An endangered breed of landrace heritage cattle, which means that the breed formed under local conditions

for local purposes – usually in isolation. The breed is descended, along with the Florida Cracker, Corriente and Texas Longhorn, from the same original Spanish stock left behind along the Atlantic and Gulf Coasts of America by the Spanish conquistadors in the early 1500s. Of the Spanish breeding stock of small, hardy cattle, which were able to survive the sea crossing to the New World, some were released deliberately, in the hope that their natural instincts would allow them to survive and reproduce.

In time, these Spanish cattle acquired different names taken from the localities in which they were concentrated. The name Pineywoods was derived from a location in southern Mississippi, and it was here that the Pineywoods bred without human interference. Thus the Pineywoods breed survived for the first 350 years, adapting to its new home in the wild. The cattle that moved west into Texas evolved to accommodate a plains habitat, and developed wide sets of horns characteristic of the longhorn breeds, while those remaining in Florida, Georgia, Alabama and Mississippi had to survive in thick woods and areas of brush, in an environment that favoured small, nimble animals with slender horns that allowed passage through narrow, brushy trails. Since the mid 1800s, Pineywoods have tended to live in semi-wild conditions on very large family ranches along the Gulf Coast.

Early settlers, such as Crackers (white settlers) and Native Americans used the cattle as oxen and for meat, milk, hides, and as a trading commodity. During the early 1800s the Choctaw Indians began to migrate west in search of agricultural opportunities and brought a limited number of livestock, including Pineywoods cattle, with them. Many people and livestock were lost due to the harsh travelling conditions before reaching Oklahoma, therefore, the majority of Spanish-type livestock must have been introduced to Oklahoma prior to the 1830s.

Pineywoods numbers began a decline in the late 1800s and early 1900s as they were displaced by improved English and European cattle in the south-eastern United States. As the overall popularity and abundance of the Pineywoods declined, only a few families continued to keep purebred herds. During this time the agricultural programmes of the land grant universities were promoting highly- bred domestic cattle and saw these as inferior 'scrub' animals. The effect, therefore, was to endanger the Pineywoods' existence as a breed. As time passed, these herds became isolated from one another to the point when each became a unique and self-contained strain.

The Pineywoods breed has been developed largely through natural selection. They developed natural resistance to most of the diseases, insects and parasites of the region and were able to forage on rough vegetation that commercial cattle would not touch. Even though they graze grass, like domestic cattle, Pineywoods will also eat brush and tree leaves and twigs like goats, making more efficient use of the land.

The Pineywoods often resembles the related Texas Longhorn and Florida Cracker cattle in colour, being generally red, brown, or occasionally black-and-white, spotted or speckled. Compared with the Texas Longhorn, the horns of Pineywoods cattle are small to medium in length and tend to curve inwards or upwards, allowing them to ward off predators. Mature weight ranges from 600–1,000lbs (270–454kg) and occasionally more depending on the environment. The smaller structure and horn size has been retained to meet the needs of the farmers and loggers of southern Mississippi. Despite their apparent advantages, at least in some regions, the term 'pineywoods' has come to mean a thin, bony, or poor-looking cow.

In 1999, some estimates were that the herd had shrunk to fewer than 200 breeding animals. The Pineywoods Cattle Registry & Breeders Association was formed to preserve the breed, and members are dedicated to saving it, viewing it as a national resource and attempting to keep it in natural conditions. Pineywoods are on the 'critical' list of the American Livestock Breeds Conservancy, its definition of critical being fewer than 200 North American annual registrations and an estimate of fewer than 2,000 in the global population. A population amounting to fewer than 1,000 head of pure stock, but not necessarily registered, has been located by the Pineywoods Cattle Registry.

PINZGAUER

A dual-purpose breed from the Pinzgau region of the Salzburg province of Austria. It is thought that the ancestors of Pinzgauer cattle were introduced into their primary breeding area in the Hohe Tauern mountains by the Celts in around 800 BC. Genetically, the Pinzgauer is most closely related to North German lowland breeds.

The Pinzgauer was first mentioned as a breed in 1846. It was originally bred for milk, beef, and draught work for use on farms, in breweries, and in fields where sugar-beet was grown. The Pinzgauer became the most popular cattle breed in Austria-Hungary, subsequently expanding throughout Eastern Europe. By December 1890, the Pinzgauer population had grown to 101,880 in Bavaria, but eventually collapsed as a result of industrialization after the First World War. Demand for the cattle decreased, and the breed was replaced by better milk-producing breeds such as Fleckviehs. The Bavarian Pinzgauer Cattle Breeding Association was founded in 1896. By 1930, Bavaria had only 85,000 Pingzauer cattle.

The Pinzgauer breed includes a naturally polled type, known as Jochberg Cattle. These descended from a single, almost totally white calf that was born in the Tyrol in 1834. They were considered crippled and useless because they could not be yoked, but now that this is no longer necessary, the hornless breed is best adapted to modern husbandry. There are now fewer than 50 hornless Pinzgauer cattle in the world, so the type is endangered. Since 1988, moreover, there have been only two hornless bulls at the insemination station near Salzburg.

Pinzgauers are distinctively marked with chestnut-brown sides and white back and underside. Black animals have occurred, but rarely, and they were once frowned upon. After 1900, black bulls were removed from the breeding system, and the black colour vanished. All Pinzgauers have the breed-typical white pattern in common, this being a broad white stripe lengthwise along the whole back, while the abdomen, chest, udder, and tail are also white. Males range in weight from 2,200–2,425lbs (1000–1100kg), while the females come in at 1,320–1,540lbs (600–700kg).

The Pinzgauer is regarded as an endangered breed, with the population decreasing by about 10 per cent per year. The American Pinzgauer Association has, as its primary objective, the development, registration and promotion of the Pinzgauer breed of cattle in the USA.

RED POLL

A dual-purpose breed of cattle developed in England in the latter half of the 19th century. The Red Poll is derived from the original cattle of Norfolk and Suffolk, thought to have been introduced by the Romans. The horned Norfolk, which was a beef-type cattle, frequently blood-red in colour, was crossed with the Suffolk polled bull from an excellent dairy breed of predominantly dun-coloured cattle. The polled gene in the Suffolk suppressed the Norfolk horn, and in 1863 the name Norfolk and Suffolk Red Polled cattle was adopted, the first standard description having been agreed in 1873. The first herd book followed in 1874, and in 1883 the breed became known as 'Red Polled'. The Red Poll Cattle Society was formed in 1888. The colour of the breed was by then established as preferably deep red, with white touches only on the tail switch and udder. Red Poll cattle are mainly used as beef sucklers, although there are a few dairy herds in England. They are known for easy calving and for successfully rearing a high proportion of their calves. They do well on poor soils that are generally lacking in fertility.

The breed was introduced into the United States by G.P. Taber of New York State in 1873. It is the oldest registered breed in the United States, and is beginning to be re-introduced into some areas of the western US for the small ranch and backyard farmers, in that its docile nature makes it easy for the novice to rear.

Red Poll cattle were imported into Australia in the mid 1800s where they are now used for beef production. The first identified breeder in Australia was James Graves, in around 1870, although there is evidence of earlier herds.

The Red Poll breed was first brought to New Zealand in 1898, but a herd was not established until 1917, when 22 animals were transported from Australia.

OPPOSITE & BELOW: Once one of the dominant English dairy breeds, the Red Poll has also retained its dual-purpose characteristics.

ROMAGNOLA

The breed derives from *Bos primigenius podolicus*, a wild ox that once inhabited the Italian peninsula, and to a greater extent from *Bos primigenius nomadicus*, a bovine originating in the Euro-Asian steppes, which came to Italy during the fourth century AD with the Gothic invasion led by Aginulf. Therefore, the Romagnola combines the characteristics of both major types of aurochs, the ancient wild cattle which were the forebears of the modern *Bos taurus* and *Bos indicus* cattle breeds.

Romagnolas are bred primarily for beef, and were often used as draught beasts in the past. They are white- or grey-coated with a black pigmented skin and have upward-curving horns. Calves are born red-coated but soon change to grey. Adult weights are on average 2,750lbs (1250kg) for males and 1,650lbs (750kg) for females. Disposition is good, and rapid weight gain, economical feed conversion and good-quality meat are claimed for the breed.

Some animals were exported to Scotland in the early 1970s and today the breed is present in Great Britain, Ireland, North and South America, Australia, New Zealand and Africa.

ABOVE, OPPOSITE & PAGE 98: Salers cattle originated in France's Massif Central.

SALERS

The breed originated in the Cantal, a department of the Auvergne in the Massif Central of France, where there is a harsh and variable climate and altitudes of 2,000–6,000ft (610–1830m). The Salers is one of the oldest breeds in the world, with prehistoric cave paintings suggesting that a similar type inhabited the area for 7–10,000 years. Salers appear to be closely related to the old Celtic and African breeds, and were probably located in the Massif Central when red cattle migrated from Africa through the Iberian Peninsula and on into northern Europe and the British Isles. In the 19th century, Ernest Tyssandier d'Escous began to improve the race theough selective breeding.

Originally bred for work, the Salers was also appreciated for its ability to withstand extreme variations in temperature, for its fertility, for its ease of breeding and for its milk and meat. In Cantal, transhumance is practised, whereby cattle are moved to high mountain pastures in summer, then brought down to lowland areas where they spend the winter.

The Salers is a large breed, the female weighing in at between 1,540–1,650lbs (700–750kg) and standing 4.6ft (1.4m) tall. They have thick mahogany red or black coats and long, lyre-shaped, light-coloured horns. A small percentage are born without horns (polled). Females can produce almost 6,615lbs (3000kg) of

fat-rich milk each year of her life. The milk is traditionally used to produce *Appellation d'Origine Contrôlée* (AOC) cheeses such as Cantal and Salers. The Salers is also used to produce veal calves by crossbreeding with Charolais.

The Santa Gertrudis is the first beef breed to have been developed in the USA.

SANTA GERTRUDIS
A tropical beef breed of cattle developed in southern Texas on the King Ranch, and named for the Spanish land grant where Captain Richard King originally established the ranch. The Santa Gertrudis was officially recognized by the United States Department of Agriculture in 1940, becoming the first beef breed formed in the United States, by mating Brahman bulls with Shorthorn cows, the final composition

being about three-eighths Brahman and five-eighths Shorthorn. In 1950 the Santa Gertrudis Breeders International Association was formed at Kingsville, Texas.

In 1918 the King Ranch purchased 52 bulls of three-quarters to seven-eighths *Bos indicus* breeding to mate with 2,500 pure-bred Shorthorn cows on the ranch. At this time the American Brahman breed did not exist as such, nor were there purebreed *Bos indicus* available in the United States. Monkey was born in 1920, a son of Vinotero, one of the bulls purchased in 1918, and which became the foundation sire for the breed. With the birth of Monkey, and a decision to line-breed, came a very uniform and hearty breed of beef cattle. These cattle are red in colour, display a blend of *Bos indicus* and *Bos taurus* attributes, and may be polled or horned. In addition to being a hardy breed, other characteristics include good milking ability, good beef production, excellent mothering ability, ease of calving, high heat tolerance and resistance to parasites, and an ability to turn off (sell or use for food) a steer at just about any age. The steers also show good weights for their age as well as good weight gains whether on pasture or in feedlots.

Santa Gertrudis cattle are known the world over for their ability to adapt to harsh climates. They were exported to Australia in around 1951 and have been subjected to inspection and classification since that time. The Santa Gertrudis Breeders (Australia) Association was established in 1954 and the Santa Gertrudis Group Breedplan has operated in Australia since 1994. Anna Creek, Australia's largest cattle station raises Santa Gertrudis cattle. There are 11,500 registered in the United States.

SENEPOL

The Senepol breed of beef cattle was developed on the Caribbean Island of St. Croix from N'Dama cattle, imported in the late 19th century, and crossings with Red Polls. The Senepol breed combines the N'Dama characteristics of heat tolerance and insect resistance with the docile nature, good meat, and high milk-production of the Red Poll. Senepols are polled, short-haired, and coloured red, black or brown.

N'Dama cattle were imported from Senegal to St. Croix in the 19th century, being better suited to the climatic conditions than European breeds. One of the largest herds of over 250 head was owned by Henry C. Nelthropp at the Granard Estates. In 1918 Henry's son, Bromley, bought a Red Poll bull from Trinadad to improve the cows' milking ability and remove their long horns. More such bulls were used in the following years and the cattle were selected for a solid red colour, natural polling and heat tolerance. These offspring were dispersed to four main herds on the island.

The name Senepol was adopted in 1954 and a breed registry was established in the late 1960s. Assisted by the United States Department of Agriculture, the College of the Virgin Islands Extension Service began a programme of farm performance testing in 1976. In 1977 22 cows were taken to the United States and the breed has since spread across the southern states.

There are now more than 500 breeders with more than 14,000 registered cattle. Senepols are also to be found in Australia, Venezuela, Mexico, the Philippines, Zimbabwe and Brazil.

SHORTHORN & MILKING SHORTHORN

The Shorthorn is possibly the most famous and influential breed of cattle in the history of agriculture. It was among the first livestock breeds to be improved, during the 1700s, and had one of the first herd books, established in 1822. Since the early 1800s and until recently, Shorthorns were the most popular cattle in Britain, and they were exported around the world. The story of the Shorthorn has been recorded in countless publications, and images of red, white and roan Shorthorns dominate sporting art. Yet the breed is now in decline, and its rise and fall reflects the great changes that have occurred in agriculture over the past two centuries.

The Shorthorn was historically the Durham, because it originated in the county of Durham in England. Imported Dutch cattle were crossed with native stock and were selected for performance in both meat and milk production. The breed became

The Shorthorn played a large part in the development of modern cattle breeds. Today, sadly, its numbers are in decline.

known early in the 1800s, especially through a travelling promotional exhibit of the famous 'Durham Ox', calved in 1796, which weighed over 3,500lbs (1585kg) at ten years old, and which brought much acclaim to the breed. Shorthorns were first imported to America in the late 1700s, the largest number of cattle having been brought in after 1820. The breed was initially concentrated in Ohio and Kentucky, a region rich in grass and corn, but by the end of the 1800s it had spread throughout America. The Shorthorn was valued

for its dairy and beef qualities and was also used as a draught animal.

Early in the 1900s, the breed was formally split into a beef type, known as the Beef Shorthorn or simply Shorthorn, and a dairy type or Milking Shorthorn. Most breeders favoured selection for beef, and this trend has continued, especially with the rise of the Holstein as the dominant dairy breed. The Milking Shorthorn, despite its many fine qualities, could not compete with the quantity of milk produced by the Holstein, and the breed lost favour as a result.

Another factor in the decline of the Milking Shorthorn has occurred more recently. In an effort to increase milk production, the breed's herd book has been opened to substantial introductions of outside blood, first from the Illawara (Australian Shorthorn) and then red-and-white Holsteins. Today, many of the bulls registered as Milking Shorthorns are actually one half or more Holstein. While these introductions did increase production, they also reduced the breed's genetic uniqueness, making it less distinct from the Holstein and therefore less useful for commercial crossings. It also lost consistency of performance in its historic traits, such as the ability to produce on grass, which is a major obstacle in its promotion for low-input dairying.

The Milking Shorthorn is medium to large in size, with females weighing 1,200–1,400lbs (540–635kg) and bulls about one ton. Milking Shorthorns are red, white, roan, or a mixture of the three, sometimes with extensive speckling. Most cattle are horned.

Although there are several thousand Milking Shorthorns (Dairy Shorthorns) in Britain, the breed is declining both globally and in North America, although pure American strains are the conservation priority in the United States.

SIMMENTAL

A versatile breed of cattle originating in the valleys of the Simme river, in the Bernese Oberland of western Switzerland. One of the oldest and most widely distributed of all the breeds of cattle in the world, and recorded since the Middle Ages, the Simmental has contributed to the creation of several other famous European breeds, including the Montbéliarde (France), the Razzeta d'Oropa (Italy) and the Fleckvieh (Germany).

Historically, the Simmental has been used for dairy, beef and as a draught animal, being particularly famous for the rapid growth of its young, if given sufficient feed. Its traditional coloration has been variously described as 'red and white spotted' or 'gold and white', although there is no specific standard, and the dominant shade varies from a pale yellow-gold to a very dark red (the latter being particularly popular in the United States). The face is normally white, and this characteristic is usually passed to crossbred calves, this being genetically distinct from the white face of the Hereford.

OPPOSITE, BELOW & PAGE 104: The Simmental has contributed to the creation of several other famous European breeds.

SOUTH DEVON

Also known as Orange Elephants, South Devons are the largest of the British native breeds. They are believed to be descendants of the large red cattle of Normandy, which were imported during the Norman invasion of England in 1066. The breed is a rich, medium- red with copper tints, though it varies in shade and can even appear slightly mottled. The breed today is predominantly used for beef production although it has been milked in the past.

The South Devon of today originated in England's South-West, in an area of Devon known as the South Hams, from where they spread right across the counties of Devon and Cornwall. Historical evidence indicates that isolation caused the divergence of the North and South Devon cattle into physically distinct types, though occasional crossings between the two occurred until the mid-19th century.

By 1800, the South Devon had become established as a breed. They were powerfully built and supplied rich milk and good beef that was

finely grained and marbled, and they were also used to pull ploughshares until well into the 19th century.

Careful selection of breeding stock improved the breed considerably. The official governing body, the South Devon Herd Book Society, was founded in 1891, when it was recognized by the government as such, and the South Devons become one of the 14 breeds of cattle whose herd books date back to the second half of the 19th century.

During the early years of the 20th century the South Devon was regarded as a dual-purpose breed, and although most herds were milked during and soon after the Second World War, the trend has been increasingly towards producing beefier sires since the 1960s. Although it is now a purely beef breed, the dual-purpose heritage has significance for the suckling of calves.

OPPOSITE & BELOW: The South Devon is the largest of the British native breeds, possibly descended from cattle brought to England during the French invasion of 1066.

Recently, there have been attempts to re-introduce the breed into dairy farming. Taverner's Farm in Devon has a small herd which it uses to produce its own brand of Orange Elephant ice cream. It has also attempted to produce a bull for this herd by artificial insemination of its aged prize heifer with 45-year-old semen from the Rare Breeds Survival Trust's national genetic archive. As of August 2008, however, this has proved unsuccessful.

The breed is exceptionally adaptable to varying climatic conditions and is presently well-established on five continents. South Devons were one of the few British breeds to have been selected for drought purposes as well as for beef and milk. The first importations into Australia were of milking cows carried on sailing ships. Several large importations occurred in the late 1800s and early 1900s, but the breed then lost its identity through crossbreeding. Importations from Britain took place in 1969, and the first purebred animals were imported from New Zealand in 1971. The breed occurs in most states of Australia.

The first South Devons were taken to the United States in 1969, and in 1974 the North American South Devon Association was formed for the purpose of development, registration and promotion of the breed in that country.

OPPOSITE & ABOVE: The Tarentaise can exist in high altitudes where there is difficult terrain.

TARENTAISE

Descended from the domestic cattle of the Tarentaise Valley in the French Alps, where they were isolated from other breeds for many thousands of years. The Tarentaise adapted in such a way as to allow them to exist in high altitudes and be able to range in very steep and rough terrain to forage.

Today, they are used to produce a cheese with a distinct flavour. These cattle have found special niches around the world, and are used primarily in the USA for producing crossbreds suited for tough rangeland conditions and higher elevations. They are also bred for the beauty of their markings and for their docile demeanor.

TEXAS LONGHORN

The first domestic cattle to reach the Americas were imported from Spain by the conquistadors from about 1493 and spread out to form a hardy race adapted to the local conditions, becoming semi-wild in some places. It is likely, but not certain, that cattle imported by settlers from the British Isles contributed to the stock after about 1600, and more certainly from the 1820s and 30s. The Texas Longhorn is used to add hybrid vigour when crossed with other breeds.

It is known for its lean beef, which is lower in fat, cholesterol and calories than that of other breeds. There was a time, however, when leaner beef was not as attractive in an era where fat meat was highly prized, and the Longhorn's ability to survive, often on the poor vegetation of the open range, was no longer such an important issue. Other breeds demonstrated traits more highly valued by the modern rancher, such as the ability to put on weight quickly. The Texas Longhorn stock slowly dwindled until, in 1927, the breed was saved from almost certain extinction by enthusiasts from the United States Forest Service, who collected a small herd of stock to breed on the Wichita Mountains Wildlife Refuge in Lawton, Oklahoma.

A few years later, J. Frank Dobie and others gathered small herds to keep in Texas state parks. They were kept largely as curiosities, but the stock's many positive qualities rekindled interest in the breed. Today, the breed is still used for beef, although many Texas ranchers keep it purely because of its importance in the history of the state and its connections with the Old West.

In other parts of North America the breed is more thoroughly utilized. Longhorn cattle have a strong survival instinct and can find food and shelter even in adverse conditions. Longhorn calves are very

LEFT & OPPOSITE: Texas Longhorns are medium-sized, fairly lean cattle with large, impressive horns.

LEFT: Texas Longhorn cows calve easily and often hide their calves from predators.

OPPOSITE: Wagyu produce the enormously expensive Kobe beef.

tough and can stand up sooner after birth than other breeds, and there have been cows that have bred for up to 30 years. Some ranchers keep Longhorns for their easy calving, in that a cow will often go off on her own to a safe place to have her calf before bringing it home. They are also known to hide their calves in safe places to avoid predators, sometimes causing difficulty for ranchers, who may need to work on the animal.

This is a breed of cattle known for its characteristic horns which, in exceptional animals can be as much as 7ft (2.1m), measured from tip to tip in a straight line. The horns can also have a slight upward tilt or even a triple twist at the tips. Texas Longhorns are known for their diverse colouring. The Texas Longhorn Breeders Association of America and the International Texas Longhorn Association serve as the recognized registries for the breed.

Texas Longhorns with superior genetics can often fetch $40,000 or more at auction with the record of $160,000 having been set in recent times. Due to their gentle disposition and intelligence, Texas Longhorns are being trained increasingly as riding steers.

The Cattlemen's Texas Longhorn Registry is dedicated to preserving the purest Texas Longhorn bloodlines. Using visual inspection of cattle by the most knowledgeable Texas Longhorn breeders and the use of bloodtype analysis to further identify parentage, CTLR has the ideal of preserving full-blood Texas Longhorn cattle for posterity.

TULI

Migrating African tribes, moving south from East Africa, brought with them humped cattle. During many years, animals displaying bad temperaments and which were also poor breeders were culled, resulting in a docile, fertile breed. This selection has supplied a basis of good genetic material with which to breed for present-day requirements.

The Tuli is a medium-sized, pure African Sanga breed which, in its neat, compact frame, incorporates high fertility, hardiness, adaptability and excellent beef qualities. Its sleek and glossy short-haired coat varies in colour from silver through to golden-brown to rich red.

The name Tuli stems from the Ndebele word *utulili*, meaning dust, and vividly conveys the arid environment from which the Tuli stemmed. It is related to the Tswana breed from Botswana, and Tuli cattle have been exported to Argentina, Mexico and the United States In South Africa a composite of Tuli and Limousin cattle has recently been developed named Tulim cattle.

WAGYU

Wagyu refers to cattle that are genetically predisposed to produce beef with a high percentage of oleaginous unsaturated fat in the marbling. There are five major breeds of Wagyu cattle (*wa* meaning 'Japanese' and *gyu* 'cow'): Japanese

THIS PAGE & OPPOSITE: Wagyu are either red or black in colour.

was the Shogun who discovered that soldiers grew bigger and stronger if they ate meat. Consequently, meat was at first confined only to the emperor and then to the elite at the imperial court. The cattle destined for the emperor's table were naturally very pampered. Later, soldiers took their eating habits with them to the villages and influenced Japanese cuisine.

Grazing areas are only available on mountain sides and in other

Black, Japanese Brown, Japanese Polled, Japanese Shorthorn, and Kumamoto Red. The meat is known worldwide for its enhanced eating quality, tenderness and juiciness, and thus commands a high market value. The meat also contains a higher percentage of Omega-3 and Omega-6 fatty acids than typical beef. It is best-known to foreigners as Kobe, Matsuzaka or Omi beef.

Originally Japan was mainly vegetarian, the consumption of meat being banned on religious grounds. Cattle were used only as beasts of burden in the production of rice. It

remote areas where rice and other crops cannot be grown. Moreover, as each Wagyu is extremely valuable, and the herds are so small, owners will not risk the physical dangers associated with grazing, preferring to keep animals penned instead. Because of this, different breeding and feeding techniques have had to be used, such as massage or adding beer or sake to feeding regimes; massage may have been used to prevent muscle-cramping in conditions where animals had insufficient room to stretch their legs.

WELSH BLACK

Commercial exploitation of the Welsh Black breed meant that drovers would direct them to English markets on foot. Herds from south-west Wales travelled towards Hereford and Gloucester up the Tywi Valley to Llandovery, while herds from South Cardiganshire reached Llandovery through Llanybydder and Llansawel. Drovers would then return to Wales with large amounts of money, which made them the targets of bandits and highwaymen. The result was the formation, in 1799, of the Banc yr Eidon (Bank of the Black Ox) in Llandovery, which was later purchased by Lloyds Bank.

By the turn of the 19th century, 25,000 cattle were being exported from Wales every year. Prior to the 1960s, few cattle were seen outside the UK, but they can now be found in Canada, New Zealand, Australia and Germany as well in countries such as Saudi Arabia, Jamaica and Uganda.

As the name suggests, the cattle are naturally black. They generally have white horns with black tips, but these may be removed, and there are also naturally hornless (polled) strains. Red individuals occur occasionally, with red and other colours having been more common in the past. The Welsh Black's hardy nature, coupled with its habit of browsing as well as grazing, makes it ideal for rough pasture such as heathland and moorland, also for conservation grazing. Traditionally bred for both milk and beef, it is now used commercially only for beef.

OPPOSITE & RIGHT: The Welsh Black is very hardy and will happily browse as well as graze, making it an ideal animal for rough ground.

CHAPTER TWO
PIGS

AMERICAN LANDRACE

The various strains of Landrace swine are the descendants of the famous Landrace hogs that were developed in Denmark. The development of the breed began in about 1895 and resulted from crossing the English Large White Hog with the native swine. It was largely though the use of the Landrace that Denmark became the great bacon-exporting country, with England as its chief market.

American Landrace are descended from Danish Landrace and Large White Hogs. It was difficult to obtain the Danish Landrace because the Danish authorities placed heavy restrictions on the use of the breed in other countries. It was finally released from breeding restrictions after a petition in 1949 to make the American Landrace a relatively new breed. It was then developed by a selection and testing programme which brought it approximately to the American Landrace we know today.

The hair colour of the American Landrace must be white. Dark skin spots are considered undesirable, but a few freckles are permitted. The pig has a long body with 16 or 17 pairs of ribs. The arch of the back is much less pronounced than that of most other breeds, and on some the back is almost flat. The head is long and rather narrow and the jowl clean. The ears are large and thin and carried close to the face. The rumps are long and comparatively level and the hams are plump but trim. The sides are long, of uniform depth, and well let down in the flank. The sows are prolific and satisfactory mothers, and have always been noted for their milk-producing abilities. They usually attain their maximum milk production after five weeks of lactation, which is later than other comparable breeds.

AMERICAN YORKSHIRE

The American Yorkshire is the American version of the English Yorkshire pig (now usually known as the English Large White). It is white in colour, with erect ears, and is the most common swine breed in the United States today.

The Yorkshire breed was developed in Yorkshire, England, in around 1761. In 1830 the first Yorkshires were imported to the United States, specifically to Ohio, but because of their slow growth-rate did not become popular until the late 1940s, when there was a period of rapid breed expansion. A large percentage of Yorkshires were

OPPOSITE: The American Landrace is a cross between the Danish Landrace and the English Large White Hog. A piglet is shown here.

ABOVE: The American Yorkshire is very similar to its English cousin.

brought in from Canada, where the breed was popular, due to its ability to produce the kind of carcase demanded in that country. More Yorkshires were also being imported from England where they were known for having greater substance, ruggedness and scale. By selection and the use of the imported pigs, the American Yorkshire met the needs of the pork producer and the demands of the market in the USA. Today, Yorkshires are found in nearly every American state, with highest populations in Illinois, Indiana, Iowa, Nebraska and Ohio. The modern breed is muscular with a high proportion of lean meat. Yorkshire data, including details of growth, sow productivity, and backfat formation, is the most extensive of the documented livestock performance records in the world.

BERKSHIRE

Said to be Britain's oldest pig breed, the Berkshire originated in the Faringdon and Wantage regions of the English county of Berkshire (now in Oxfordshire). Today's animals are the descendants of the royal herd maintained at Windsor Castle 300 years ago. Berkshires apparently became popular after they were 'discovered' by Cromwell's troops while stationed at Reading during the English Civil War.

Berkshires are early-maturing black pigs that often have white markings on their legs, faces, and the tips of their tails. The snouts are dished and are of medium length, and the ears are fairly large and erect or lean slightly forward. Bershires have fine wrinkle-free necks and well-sloped shoulders. They are compact animals, with good muscling, short legs and deep bodies.

In Britain, the breed is maintained by the Rare Breeds Survival Trust, at Aldenham Country Park, Hertfordshire, and at the South of England Rare Breeds Centre in Kent. The Berkshire is listed as 'vulnerable', there being fewer than 300 breeding females in existence. In the United States, the American Berkshire Association, established in 1875, only pedigrees pigs directly imported from established English herds, or those that can be traced directly back to them. The pig is also bred in the Kagoshima Prefecture, Japan, under the trademark name Kagoshima

OPPOSITE & RIGHT: The Berkshire is probably England's oldest pig breed.

Kurobuta (black pig). In New Zealand it is estimated that there are now fewer than 100 purebred sows. In literature, the 'Empress of Blandings', in P.G. Wodehouse's Blandings Castle series, and 'Pig-Wig', a sow in Beatrix Potter's *Tale of Pigling Bland* were both Berkshire pigs.

OPPOSITE: Berkshires are early-maturing black pigs, often with some white markings.

ABOVE: The Chester White is an American breed from Pennsylvania.

CHESTER WHITE

First known as the Chester County White, but now simply as the Chester White, the breed originated in Chester County, Pennsylvania. Chester Whites are medium to large pigs with white skin (a little black spotting may sometimes appear) and floppy ears.

The development of the breed began in 1815–18, using strains of large, white pigs common to the USA's north-east and a white boar imported from Bedfordshire,

England; some believe that Chinese pigs were also added to the mix. By 1884 a breed association had been officially formed but competing organizations, sometimes devoted to individual strains, continued to appear into the early 20th century. Finally, in 1930, all breed organizations were consolidated under the Chester White Swine Record Association, which helped the spread of the breed to the rest of the US.

Today the Chester White is regarded as a versatile breed suited to both intensive and extensive husbandry. Though not as popular as the Duroc, Yorkshire, or Hampshire breeds, it is actively used in commercial crossbreeding operations. It is best known for producing large litters and for good mothering skills. Although the boars can be rather aggressive (as are most swine), the sows are very easy to manage.

CHOCTAW

A descendant of livestock brought to the Americas by Spanish conquistadors from the 1500s onwards. The Choctaw pig was used not only by Native Americans but also by European settlers and a succession of other peoples in the south-eastern United States for over 300 years.

The Choctaw people and their livestock migrated from the Deep South to Oklahoma Territory in the early 1800s. The United States government forced the Five Civilized Tribes out of Mississippi and Alabama in 1830, and they too brought more pigs to Oklahoma, and it is from these that today's animals are descended, their appearance having changed very little in 150 years.

Choctaw hogs have two distinctive characteristics indicative of their Spanish ancestors. First, their toes can be fused, forming hooves like those of a mule. Secondly, many have fleshy wattles on either side of their necks. Choctaws are black, sometimes with white markings, and at about 120lbs (54kg) are relatively small for domestic pigs. They are nevertheless quick and athletic, with heavy forequarters, being obviously built for survival.

Feral descendants of Spanish pigs are much more common than non-feral ones, but the Choctaw is a pure Spanish breed and distinct from the feral hog populations in the Choctaw National Wildlife Refuge area, which are of mixed ancestry. The breed is now reduced in population to a few hundred animals, most of them in the care of the Choctaw Nation of Oklahoma. The American Livestock Breeds Conservancy describes its status as critically rare and worthy of a high priority where conservation is concerned.

OPPOSITE & BELOW: The Duroc is an old American breed that is always red in colour.

DUROC

An older American breed that formed the basis for many mixed-breed commercial pigs. Durocs are red, large-framed, of medium length and muscular, with partially drooping ears.

In 1812, early red hogs were being bred in New York and New Jersey. They were large in size, and characteristically had large litters and the ability to grow quickly. The

foundation that formed today's Duroc breed comprised Red Durocs from New York and Jersey Reds from New Jersey. The Duroc is reputed to have been named after a famous Thoroughbred stallion of the day. Little of the actual crossing of the two strains was done in the two states in which the pigs originated, but much of the amalgamation into one breed took place in the Midwestern states. The Duroc's devotion to caring for its young makes it ideal as an outdoor pig, both in the dam or sire line, and its succulence and heavy muscling makes it very suitable for anything from light pork to heavy hog production.

GLOUCESTERSHIRE OLD SPOT
An English breed of pig that is predominantly white with black spots. It is named after the English county of Gloucestershire. The Old Spot was once a very popular breed of pig, but with the advent of intensive farming, certain pale, lean, high-yield breeds have been chosen to

suit the factory conditions and needs of mass-production, with the result that many old pig breeds have died out or are greatly diminished. Owing to consumer pressure in the United Kingdom, however, and changes in the law, both attributable to an increasing awareness of, and concern about, farming conditions, pigs have come to be increasingly reared outdoors. Also, more consumers are looking for quality meat, as opposed to cheap, bland alternatives. Therefore old breeds, such as Old Spots, being well-suited to living outdoors, are increasingly being chosen by farmers looking to add value to their product.

Old Spots are sometimes known as 'orchard pigs', being traditionally pastured in orchards where they ate windfall apples which, in small-scale orchard management, helped to keep down pests. They are also good foragers and survive well in pastures without supplementary feeding.

Often referred to as a 'bacon' pig, due to its significant depth of body that provides a larger percentage of bacon per hundredweight of carcase, Old Spots often carry more fat than more commercially popular breeds. They tend to be calm, good-natured animals, another trait that makes them desirable to small-scale and backyard farmers. The females make devoted mothers, while the males seldom pose a threat to piglets.

The Gloucestershire Old Spot is currently on the critical list of the American Livestock Breeds Conservancy, meaning there are fewer than 200 annual registrations in

OPPOSITE & RIGHT: Gloucestershire Old Spots were often used to clear orchards of windfall apples.

the United States and fewer than 2,000 in the estimated global population. In the UK, the Old Spot is listed as 'Category 5, Minority' by the Rare Breeds Survival Trust, there being fewer than 1,000 registered breeding females in existence. In Britain, an application has been made to the European Commission to gain Traditional Speciality Guaranteed status for Old Spot pork.

GUINEA HOG

Also called the Pineywoods Guinea, Guinea Forest Hog, Acorn-Eater and Yard Pig, the Guinea Hog is a breed developed in the United States, the name deriving from its origins on the Guinea coast of West Africa. The true African Guinea Hog is a large, red breed with upright ears, bristly hair and long tail, suggesting genetic influences from the Nigerian Black or Ashanti pig. In about 1804 Thomas Jefferson acquired some of these pigs, which had arrived from Africa via the Canary Isles. The original strain, although basically black, also had a hint of red in their coats and were consequently called Red Guineas; that strain, well-known at the beginning of the 19th century, is extinct.

The Guineas were later crossed with other breeds, including Appalachian English and Essex pigs, and West African Dwarfs. This new breed, the American Guinea Hog retained its black colour but lost the red tint and is sometimes called the Black Guinea. These pigs were popular with subsistence farmers, not only because of their ability to forage for themselves, but also because their habit of eating snakes made the farmyard safe for children and other livestock.

The breed fell out of favour after around 1880 and for a while was in

ABOVE & OPPOSITE: The Guinea Hog was developed in American from the African Guinea Hog and other breeds.

danger of being entirely lost. Although the Red Guinea no longer exists, its exact relationship with the American Guinea, and what proportions of other breeds are in its background, are not known for certain. That there is a relationship, however, is shown by the occasional birth of a reddish pig to the normally bluish-black American Guinea parents. It is suspected that there were a number of distinct American

Guineas in the past. In 2005 the American Guinea Hog Association was formed which now ensures the breed's continued existence.

Guineas are one of the smaller breeds of pig, ranging from 150–250lbs (68–113kg) at maturity, and with a carcase of 50–100lbs (23–45kg), which makes them unsuitable for commercial farming. Their docile disposition makes them excellent homestead pigs, in that they are able to forage for much of their diet, which includes nuts, rodents, grass and roots, besides snakes. There are two types of Guinea Hog in North America, small-boned and large-boned, the latter having longer legs, and there is a further type in South America.

HAMPSHIRE

A breed characterized by erect ears and a black body with a whitish band around the middle, which also encompasses the front legs. The Hampshire was developed in the United States and is now one of the world's most important breeds.

The American National Swine Registry notes that this is the third most recorded breed of pig in the

United States, and probably the oldest American breed. Some of the first to arrive were known as McKay hogs, because a man of that name was thought to have imported them to America. Importations of the breed were made from Hampshire in England between 1827 and 1839. Pigs remaining in this part of England developed into Wessex Saddlebacks, a similarly-coloured pig but with floppy ears.

From the time of its arrival in the USA, the pig had also been known as the 'Thin Rind' breed, due to the abundance of lean meat it produced. At a meeting of American breeders in 1890, however, it was renamed the Hampshire for the fact that the original pigs had come from a farm in Hampshire, England. A breed society was established at the same time and herd-book recordings can be traced back for more than 100 years.

Hampshires are used extensively as sires in the crossbreeding of pigs for pork in the USA and many other countries. They have the reputation of being the leanest of the North American breeds, where the majority of carcase competitions are won by Hampshires and Hampshire crosses. The sows of the breed have been praised for their milk capacity and longevity as mothers.

The Hampshire is an American breed developed from English stock.

HEREFORD

A medium-sized breed unique to the United States. The Hereford was developed in Iowa and Nebraska during the 1920s from Duroc, Chester White and Poland China bloodlines. Additional breeding and selection led to the identification of 100 animals as foundation stock in 1934, and the National Hereford Hog Record was formed the same year to promote the new breed, attracting 450 members within the first decade of its history.

The name 'Hereford' was inspired by the pigs' strikingly beautiful colour pattern of intense red with white trim, which was strongly reminiscent of Hereford cattle. The breed description stipulates that hogs be primarily red, with white faces and two or more white feet. The shade of red can vary, although a deep red is preferred. Hereford cattlemen were so keen on the new breed of swine that the Polled Hereford Cattle Registry Association sponsored the formation of the National Hereford Hog Record.

Herefords are adaptable and thrive both in outdoor situations and in confinement. They also do well in a wide variety of climates, and are known for their quiet and docile dispositions. Early maturation is a feature of the breed, and Hereford hogs weigh 200–250lbs (90–113kg) by five to six months of age, besides which they are easy to pasture but also grain-efficient, reaching market weight on less feed than many other breeds. Mature boars weigh about 800lbs (363kg) with mature sows coming in at about 600lbs (272kg). The sows produce and wean large litters, and make excellent, attentive mothers.

The Hereford began to decline in numbers during the 1960s with the shift away from the commercial use of purebred hogs towards a three-way cross of the Duroc, Hampshire, and Yorkshire breeds. Today, the breed population is estimated at fewer than 2,000 in the United States, most of them in the upper Midwest and Plains states. The characteristics of the

Hereford, however, make it a natural choice for a variety of small-scale production systems and, if given the opportunity, the breed should be able to earn its place in the future.

LACOMBE

Developed at the Lacombe Research Station, work on the new Canadian breed began in 1947 and was completed in 1958. The brainchild of Jack Stothart and geneticist Howard Fredeen, the Lacombe is an amalgamation of Danish Landrace, British Berkshire and Chester White pigs from the USA. The three component breeds were crossed and back-crossed to form a pure composite breed, with selection based on growth-rate, litter-size and carcase-quality of three to four randomly-selected litter mates. Any pigs exhibiting below-average performance or had coloured skin were culled and these criteria led to 76 per cent of all potential breeding stock being discarded. The resulting Lacombe had 55 per cent Landrace, 23 per cent Berkshire and 22 per cent Chester White genes.

The breed was released in 1957. Contract multiplier herds were then established, complete pedigree records were submitted to the Canadian National Livestock Records, and genetic purity for the white colour was established.

Stothart and Fredeen developed a system of selective registration, the Lacombe being the only breed to have this requirement, which provided proof of the pigs' performance potential. Breeding groups of Lacombes were released each year from 1958 onwards, and 462 breeding females were in the hands of private breeders by 1960.

Results obtained in four years of field-testing showed that the Lacombe produced carcases equivalent in merit to the Yorkshire pig, and had a 10 per cent advantage in weaning weight and post-weaning growth-rate. The only disadvantage of the breed seems to be its poor vision due to its floppy ears.

Whatever its advantages and disadvantages, the Lacombe has sadly been eclipsed by the advent of large-scale breeding companies and a focus on fewer pure breeds. It is believed that many factors, including opposition to a new hybrid breed among 'slaves to the old tradition of the purebred system' has led to the gradual decline of the Canadian-developed breed.

LARGE BLACK

Native to Great Britain, the Large Black is descended from the Old English Hog established in the 16th and 17th centuries. By the late 1880s there were two distinct types of the breed, one in East Anglia, the other in Devon and Cornwall, although the founding of the Large Black Pig Society in 1889 led to an increase in the exchange of stock between

OPPOSITE & BELOW: Large Blacks are an old British breed, there being several small herds dotted throughout the UK, but they are not farmed on a large scale.

breeders in the two regions. Large Blacks had been widely distributed throughout the country by the early part of the 20th century, and were frequently crossed with Large Whites and Middle Whites to produce bacon and pork pigs. The Large Black was also very successful in the show ring at this time; at Smithfield in 1919, the Supreme Championship was awarded to a Large Black sow that subsequently sold for 700 guineas. The same year the breed outnumbered all other breeds at the Royal Show when 121 Large Blacks were exhibited.

A change in demand from the meat trade and a developing prejudice against coloured pigs led to a severe decline in numbers throughout the 1960s. Today Large Blacks can be found throughout the British Isles, mainly in small herds, some of which were established well before the Second World War.

The sows are known to be excellent mothers with exceptional milking abilities. They have large litters of eight to ten piglets, and are able to rear them on simple rations.

This is a large breed, and the only British pig which is wholly black. Large Blacks are present in a number of countries besides the UK, including Australia, Canada and the United States, where the breed is also called the Large Black Hog. The American Livestock Breeds Conservancy lists the future of the Large Black as 'critical'.

MULEFOOT

A breed named for its solid, non-cloven hooves reminiscent of those of a mule. The pigs are typically black, on rare occasions having white spots, and typically have a weight of 400–600lbs (180–270kg) by the age of two years.

Mulefoots are most likely to be the descendants of the pigs brought to the Gulf Coast by the Spanish; but how it developed syndactylism is not clear. An apparent benefit of such animals is the elimination of hoof rot, which affects the area of the hoof between the toes.

The breed was flourishing during the early half of the 20th century but, by 1985, only one herd of

Mulefoot hogs remained, belonging to R.M. Holliday of Louisiana, Missouri. At present, Mulefoots are considered by the American Livestock Breeds Conservancy to be critically rare, as there are now fewer than 150 documented purebred Mulefoot pigs in existence.

OSSABAW ISLAND

As the Spanish conquistadors explored the coast of the Americas in the 1500s, livestock, such as pigs, were often left behind as a future source of food, and this was the origin of the Ossabaw breed. Over the following hundreds of years, the population of these feral pigs remained isolated on Ossabaw, which is one of the Georgia Sea Islands, these being barrier islands off the Georgia coast, and where they were

OPPOSITE & BELOW: Ossabaws have long snouts, upright ears, and bristly coats.

isolated from other domestic breeds. Since 1978 the island has been owned by the state and managed by the Georgia Department of Natural Resources (DNR) as a preserve.

The human population of the island was never high, and the pigs generally ranged freely over its entire acreage. Like feral pigs elsewhere, those on Ossabaw Island have had an adverse effect on natural habitats, and they are threatening endangered species, such as the loggerhead sea turtle and snowy plover, disturbing nests and eating eggs. This, plus the various other impacts they have on the ecosystem have led the Georgia DNR to recommend the eradication of all feral swine via trapping, shooting and hunting by the public. This is obviously leading to their endangerment as a species.

Ossabaws, however, are also recognized as a unique genetic resource by scientists and breed conservationists. They are the only US breed whose genetics can be traced directly back to the Iberian pigs brought to North America by the Spanish. A very small breeding population of Ossabaws are kept off the island by farmers, who market them as heritage pork and ham (the meat being dark, with a unique texture resembling the *jamón ibérico* of Spain), and there are also herds at several zoos, at Mount Vernon and at Williamsburg.

The characteristics of the Ossabaw Island breed have been shaped by the pressures of feral life in an island habitat. They are small animals, being under 20in (51cm) tall and weighing less than 200lbs (90kg) at maturity, this being partly due to the phenomenon of insular dwarfism, and individuals kept in off-island farms may grow slightly larger in successive generations. They are also hardy and very good foragers, making them useful in extensive farming (as opposed to intensive pig farming). They come in a wide range of colours, most commonly black, and there is a spotted variety. Piglets do not share the striping of wild boars. The pigs have long snouts, upright ears, and a heavy coat of bristles. Ossabaws are known to be intelligent and friendly swine in terms of temperament.

PIÉTRAIN

A breed native to Wallonia, which takes its name from Piétrain, a small village of the Walloon municipality of Jodoigne in Belgium. The Piétrain was the result of crossing English and French breeds. Before the Second World War, pigs were bred to be fat, but tastes changed and the Piétrain was selected by breeders and the Belgian pig industry in 1950 because of its double-muscled hams and thick, lean loins. In fact, it has a much higher meat-to-bone ratio than any other pig.

It was exported to France and Germany, but never made much of an impression in Britain. It is not without its problems, however: if stressed it is liable to drop dead, and its extraordinarily large hams and short legs make it unable to reproduce naturally. It is neither hardy nor vigorous. The Piétrain is found in Belgium, France and was imported into Germany in 1960–61, the main breeding areas in that country being Schleswig-Holstein, Nordrhein-Westfalen and

OPPOSITE & LEFT: The Piétrain is a Belgian breed native to Wallonia.

Wurttemberg-Baden. They are commonly used in crossbreeding in Germany to improve the quality of pork produced. The breed was recently improved by researchers at the Université de Liège.

POLAND CHINA

A pig first developed in the Miami Valley of Ohio, and deriving from many breeds. It is the oldest breed of swine in the USA. It was based on the common stock found in the area, but it is not known where that originated. It is known, however, that many settlers, that came to the region at an early date, brought with them swine from various locations. There is evidence that some of these early swine probably originated in the herd of the Duke of Bedford, who was a Berkshire breeder. Some of the stock apparently came in from Kentucky and, no doubt, was of the same breeding as the pigs which later became known as Hampshires. The early native stock varied a great deal in type and colour markings.

In 1816, the Shaker Society, through their trustee, John Wallace, secured one boar and three sows, known as Big Chinas, from a firm in Philadelphia. The boar and two of the sows were white, while the third sow had sandy to black spots. Historians believe they were the same hogs that were popular about this time in the states of Maryland, Pennsylvania and Virginia. Until 1835, the swine industry of south-western Ohio had received greater impetus, due to the beneficial effects of the Big China.

Pigs were bred for two important requirements: size and the ability to travel well, in that they were driven to market and in some cases were compelled to travel nearly 100 miles.

Today, the Poland China is recognizable as a big-framed, long-bodied, lean and muscular pig, that ranks highest in US pork production in pounds of pig per sow per year. The pigs are typically black, sometimes with white patches, and are known for their large size.

RED WATTLE

A large, red pig with a fleshy wattle attached to either side of the neck. The wattles have no known function, being a single gene characteristic usually passed to crossbred offspring. The breed comes in a variety of shades of red, some with black specks or patches, while some individuals are nearly black. The head and jowl are clean and lean, the nose is slim, and ears upright with drooping tips. The body is short-coupled and the back slightly arched. Mature animals weigh 600–800lbs (270–360kg), but may weigh as much as 1,200lbs (540kg).

The breed is known for its foraging activity, rapid growth-rate, disease-resistance and hardiness. A lean meat is produced that has been described as tender and flavourful. The sows are excellent mothers, have litters of 10–15, and provide good quantities of milk for their piglets. They have mild temperaments.

The history of the Red Wattle is not clear, but it was rediscovered in the late 1960s and early 1970s by H.C. Wengler in east Texas, and he is credited with starting the Wengler Red Waddle Hog line. (Note the spelling here is Waddle rather than Wattle; presumably Wengler wanted to make sure his pigs were unique.) About 20 years later, Robert Prentice located another herd of Red Wattle hogs in east Texas, which became the Timberline type of Red Wattles. He

also combined his Timberlines with Wengler's breed to make the Endow Farm Wattle Hogs.

At the peak of the boom in the hog market in the early 1980s there were three different registries for Red Wattles. Many people seem to have the breed but there was never a central breed association. In 1999, when the American Livestock Breeds Conservancy linvestigated the matter, there were only 42 breeding animals belonging to six breeders. The ALBC now maintains the pedigree book for the breed. Recently, a Red Wattle Hog Association was started for the betterment of the breed.

TAMWORTH

A breed originating in England, with input from Irish pigs. It is among the oldest of the pig breeds, but as with many such breeds it is not well-suited to modern production methods. It is listed as 'threatened' in the United States and 'vulnerable' in the UK by the Rare Breeds Survival Trust, there being fewer than 300 registered breeding females. The Tamworth is

The Tamworth is said to have wild boar in its bloodline, with input from native European stock.

thought to be descended from wild boars via European native pig stock. Other names for the Tamworth are Sandy Back and Tam.

The Tamworth breed originated in Sir Robert Peel's Drayton Manor Estate at Tamworth, Staffordshire, after the existing herd was interbred from 1812 with pigs known as 'Irish Grazers', that Peel had seen in Ireland in 1809. Much of the improvement of the breed took place in Staffordshire and also in the counties of Warwick, Leicester and

ABOVE, OPPOSITE & PAGE 140: Tamworths and hardy, good-natured pigs that will happily graze alongside cattle.

Northhampton. The breed appears to be among the least interbred with non-European breeds, and is therefore one of the closest to the original European forest swine.

In 1865 the Tamworth achieved English breed recognition and in 1885 the herd book for the breed was begun. Tamworths were imported into the United States by Thomas Bennett of Rossville, Illinois, in 1882. Soon they entered Canada, where a population now exists. Breed Associations for Tamworth swine are active in the UK, USA and Canada, even though the Tamworth is considered to be a minor breed.

The Tamworth has an elongated head and a long, narrow body. The ears are erect and pointed, while the face has rectilinear lines as well as the snout. Colours range from a pale-gingery to dark-mahogany red. Early in the breed history, colours were red and black, but the black coloration has been bred out. The bristle density protects the Tamworth's skin from damage from ultraviolet rays; nevertheless, when moulting occurs between June and August (in the northern hemisphere), shade is needed, along with copious mud-wallowing to prevent sunburn.

The Tamworth is a medium-sized porcine, with the full-grown boar weighing from 550–820lbs (250–370kg) and the mature sow coming in at 440–660lbs (200–300kg). It has a long neck and legs, deep sides but a narrow back. Its ham structures are muscular and firm. The breed is also known for its excellent foot structure and good skeletal system. Litter sizes are typically somewhat smaller than commercial breeds. Features regarded as undesirable are curly hair, coarse manes, turned-up noses and dark spots on the coat.

The Tamworth is known for its hardiness and toleration of adverse climates, allowing it to cope with more northerly locations, such as Scotland and Canada, where winters are severe, not only with regard to the cold but also to high winds. Tamworths graze compatibly with cattle, being able to retrieve forage that cattle leave behind; the pig is an efficient excavator when rooting for food. The breed is also used in forage-based farming systems. The sows demonstrate good maternal skills, and litters normally range from six to ten piglets. The Tamworth has an amenable disposition and enjoys human attention.

VIETNAMESE POTBELLY

A breed with 14 subspecies originating in Vietnam. Considerably smaller than standard European or American farm pigs, most adults are about the size of a medium- or large-breed dog, although their bodies are denser, weighing in at 60–300lbs (27–136kg).

Potbelly pigs can be easily distinguished from other breeds, not only by their size, upright ears and straight tail, but the pot belly and swayed back are also features of the breed, and are not indicative of obesity. Pigs of the correct weight still have the sway and belly, but the hip bones can be felt easily, with minimal pressure, and the eyes

RIGHT, PAGES 142 & 143: Vietnamese Potbelly pigs are much smaller than other breeds, the pot belly being a part of their conformation rather than the result of obesity.

(whole socket) should be quite visible. Pigs with fat rolls over their eyes, or bellies that touch the ground, are probably overweight.

Because these pigs are in the same species as ordinary farmyard pigs and wild boars, they are capable of interbreeding. The Swedish Agriculture Ministry has been assisting Vietnam with its pork production by introducing large breeds of pigs into Vietnam since the mid-1980s. Today, the Vietnamese and Swedish governments have realized that the indigenous Vietnamese pig subspecies exists only in mountainous Vietnam and Thailand. Consequently, the Vietnamese government has begun to subsidize local farmers that continue to raise the indigenous potbelly pigs because it realizes that they are neither as prolific nor as large as other breeds.

The boars, or entire male pigs, become fertile at a young age, long before they are completely physically mature. Potbelly pigs are considered fully-grown by six years of age, when the epiphyseal plates in their spines finally close.

Other miniature pigs, such as the Göttinger and the Resident of Munich are bred in medical laboratories, and are frequently kept as pets in Germany. The New Zealand Kunekune pig is significantly larger than the Vietnamese, but smaller than the commercial pig. Kunekunes are reasonably popular in the UK, having been introduced there in the early 1990s.

CHAPTER THREE
GOATS

ALPINE

Synonymous with the French-Alpine, the goat originated in the European Alps. It is a hardy animal that will thrive in any climate while maintaining good health and excellent production. A dairy breed, Alpines were brought to the United States from France and were selected for much greater uniformity, size and production than the goats that had been taken from Switzerland to France.

In the development of the breed, more importance has been given to size and production than colour pattern. Colours may range, therefore, from pure white through shades of fawn, grey, brown, black, red, buff, piebald or various shadings or combinations of all of these. Both sexes are generally shorthaired, but

RIGHT & OPPOSITE: The Alpine goat originated in the French Alps. It is very hardy.

bucks usually have a roach of long hair along the spine, and the beard is quite pronounced. The ears in the Alpine should be of medium size, fine-textured, and preferably erect. A Roman nose is undesirable.

The Alpine is more variable in size than the Swiss breeds. Mature females weigh in at around 135lbs (60kg) with males coming in at around 170lbs (75kg). The females are excellent milkers and usually have large, well-shaped udders with well-placed teats of the correct shape.

ANGORA

A breed of domestic goat that originated in Ankara (formerly known as Angora) and its environs in central Anatolia, in present-day Turkey. Angora goats were depicted on the reverse of the Turkish 50-lira banknotes of 1938–52.

The first Angoras were brought to Europe by Charles V, the Holy Roman Emperor, in about 1554, but, like later imports, were not very

OPPOSITE: Alpines are a milk breed.

RIGHT: Angora goats are valued for their fine fleece, which is known as mohair.

Angoras are more delicate that other goats and require extra nutrients in order produce good-quality mohair.

successful. Angora goats were first introduced into the United States in 1849 by Dr. James P. Davis, when he received seven adult goats as a gift from Sultan Abdülmecid I in appreciation for his advice on the raising of cotton. More goats were imported over time, until the Civil War destroyed most of the large flocks in the south. Eventually, however, the Angora began to thrive in the south-west, particularly in Texas, wherever there were sufficient grasses and shrubs to sustain them. To this day, Texas, together with Turkey and South Africa, remains the largest producer of mohair (the processed fleece taken from the Angora goat) in the US, and the second largest in the world.

A single goat produces between 11–18lbs (5–8kg) of hair per year. Angoras are shorn twice a year,

unlike sheep, which are shorn only once. For a long time, Angoras were bred for their white coats, but in 1998, the Coloured Angora Goat Breeders Association was set up to promote other colours. Now Angora goats produce white, black (deep black to greys and silver), red (the colour fades significantly as the goat gets older), and brownish fibres.

Angora goats are more susceptible to external parasites (ectoparasites) than similar animals, having denser coats. They are not prolific breeders, nor are they considered very hardy, being particularly delicate during the first few days of their lives. Angoras also have special nutritional requirements due to their rapid hair growth. A poor-quality diet will have an adverse effect on mohair development.

ARAPAWA

The most interesting group of feral milch goats remaining in New Zealand, and which were found isolated on Arapawa Island in the Marlborough Sounds. They are a relatively small breed (smaller than modern milking breeds) and come in a variety of colours, with patterns of

OPPOSITE & ABOVE: Arapawas may be of English origin, brought to their island by the early whalers, or even by Captain Cook.

white, fawn, brown and black being the most common, and some have distinctively patterned faces. The males have widely sweeping horns, while those of the females are shorter and backward-pointing.

It is widely believed that Arapawas are a surviving remnant of the Old English landrace goats which no longer exist, and possibly the descendants of a pair released by Captain James Cook in 1773.

Today they are extremely rare but, as well as being maintained in a reserve on Arapawa Island itself, a number of these goats have been removed over recent years and are now being bred by a few enthusiasts in various places throughout New Zealand. In 1993 a breeding group was exported to USA and another to Great Britain in 2004. Sadly, however, the total number of Arapawas in domestication worldwide is still fewer than 300.

BOER

Developed in South Africa in the early 1900s for meat rather than milk production. The name is derived from the Dutch word *boer*, meaning farmer. The Boer goat was probably bred from the indigenous goats of the Namaqua Bushmen and the Fooku tribes, with some crossings of Indian and European bloodlines being possible. Due to selective breeding and improvement, the Boer goat has a fast growth-rate and excellent carcase qualities, making it one of the most popular goats for meat-production in the world.

Boers have a high resistance to disease and adapt well to hot, dry, semi-desert areas. US production is centred in west-central Texas,

superior mothering skills than other breeds. Mature bucks weigh in at between 240–300lbs (110–136kg), the does at 200–220lbs (90–100kg).

Goats are browsers by nature, preferring brush, shrubs and broadleaf weeds rather than grass. Goats raised for meat production are typically raised on pasture, the main reasons for this being twofold: pastured goats are on average healthier than those that have been pen-raised; secondly, it costs far less to raise them on a diet of brush and weeds rather than on commercial feed. The ideal option is adequate

particularly in and around San Angelo. The original US breeding stock came from herds located in New Zealand, and it was only later that they were imported directly from South Africa.

Commonly, the goats have white bodies and distinctive brown heads. Like the Nubian goat, they possess long, pendulous ears. They are noted for being docile, fertile, and have

In South Africa, the Boer goat is widely regarded as the yardstick against which goats across the world are measured.

year-round grazing with only mineral supplementation. Boer goats can be raised effectively in combination with cattle or sheep due to their preference for browsing and the resulting limited impact they have on grass cover. They do compete with other browsers, however, such as deer.

An undesirable attribute is poor jaw alignment – a significant disadvantage when feeding on pasture, and unacceptable in a commercial herd sire. Good feet and legs are required: hoof rot is a common problem for goats that live

in high rainfall areas if the hooves are not clipped regularly. For breeding purposes, two well-formed, equal-sized testes in a single scrotum are desirable. One buck is normally required for every 25 to 35 does. Bucks are normally separated from the does except when breeding is specifically intended. Often does are bred for six weeks every eight months, resulting in three litters of kids every two years.

GOLDEN GUERNSEY

A rare breed from the Bailiwick of Guernsey, one of the English Channel Islands. In 1965 the Golden Guernsey was exported to Great Britain and the English Golden Guernsey Club, later to become the Golden Guernsey Goat Society, was formed.

The exact origin of these goats is uncertain, but since bones dating from 2,000 BC have been found in dolmens (megalithic tombs) on the island, it is likely that the breed began to evolve into its current form at about this time. The ancestors of the Golden Guernsey are believed to have been the Oberhasli and Syrian breeds. The first documented reference to the goat in its current form dates from 1826, when mention of a 'golden goat' was made in a guide book.

As the name suggests, the goat is golden in colour, with hues ranging from pale blond to deep bronze. They are smaller and rather more fine-boned than other British milking goats, and there is a greater variety in coat length. The males are sometimes horned but the vast majority are not. Temperaments have been described as very docile and friendly. The males, however, are said to be unusually smelly.

An efficient milking goat for its relatively small size, the Golden Guernsey produces an average yield of 7lbs (3.17kg) of milk per day; this is less than most Swiss goats produce, but the milk's high butterfat and protein content compensates for the small yield.

KIKO

Developed in New Zealand from crossings of feral goats with dairy goats in the 1980s, *kiko* being the Maori word for flesh or meat. The goats were developed for fast growth and hardiness, with minimal input needed in order for them to survive on their own. New Zealand has a large population of native goats, which roam unrestrained through the wooded hill country and mountain scrubland of both islands. They derived from the original imports of British milch goats introduced in the late 18th century to provide sustenance for whalers and sealers prior to New Zealand's colonization. Over time they have been supplemented by escaping domestic goats and farmed goats turned loose, particularly during the depressions of the 1890s and 1930s.

The Kiko was developed exclusively by the Goatex Group Limited, a New Zealand company which actively involved itself in the capture and farming of New Zealand's native goat population. All members of the company had a special interest in meat production, as a consequence of which several thousand of the most substantial and fertile native goats were allocated to a breeding programme to produce a goat with enhanced meat-production ability under natural browsing conditions.

OPPOSITE: The Golden Guernsey is possibly an ancient breed, dating back 4,000 years.

Today the Kiko is a large-framed, early-maturing goat that demonstrates exceptional conversion rates, producing lean and succulent meat. Mature bucks can weigh more than 300lbs (136kg) and mature does up to 200lbs (90kg) or more.

The American Kiko Goat Association purchased and owns the original Kiko goat registry in the US. The International Kiko Goat Association was formed in 2004 to preserve and advance the breed, appreciating its superior characteristics and dedicating itself to promote it as the best production meat goat in the world.

KINDER

A breed of dairy goat, though it can also be used for meat. The Kinder was created by crossing a Pygmy with a Nubian goat at Zedderkamm Farm, Washington state, in 1985. There are now about 1,000 Kinders registered with the Kinder Goat Breeders Association. Kinders are medium-sized goats that are well-proportioned in body length and legs. Their

RIGHT & OPPOSITE: The Kinder is the result of a crossing of a Pygmy and a Nubian goat.

compact physiques conform to dairy characteristics despite their somewhat heavy bone and lean, yet well-muscled build. On average, a mature male weighs in at 135lbs (61kg), the female coming in at 115lbs (52kg). The Kinder is a prolific, productive, alert, good-natured and gregarious breed.

LAMANCHA

A milking goat noted for its apparent paucity of external ears, although there is an auditory canal and other internal structures necessary for hearing. It is not usually possible to use ear tags or ear tattoos for identification, therefore these are normally placed on the tail web. LaMancha is medium in size, and is noted for its quiet and gentle manner.

The breed was developed in the 1930s in Oregon by Eula F. Frey, who crossed some Californian short-eared goats, believed to have Spanish bloodlines, with her outstanding Swiss and Nubian bucks. LaMancha has an excellent milking temperament and is a sturdy animal that can withstand hardship and still produce. It has established itself as a producer of high butterfat milk.

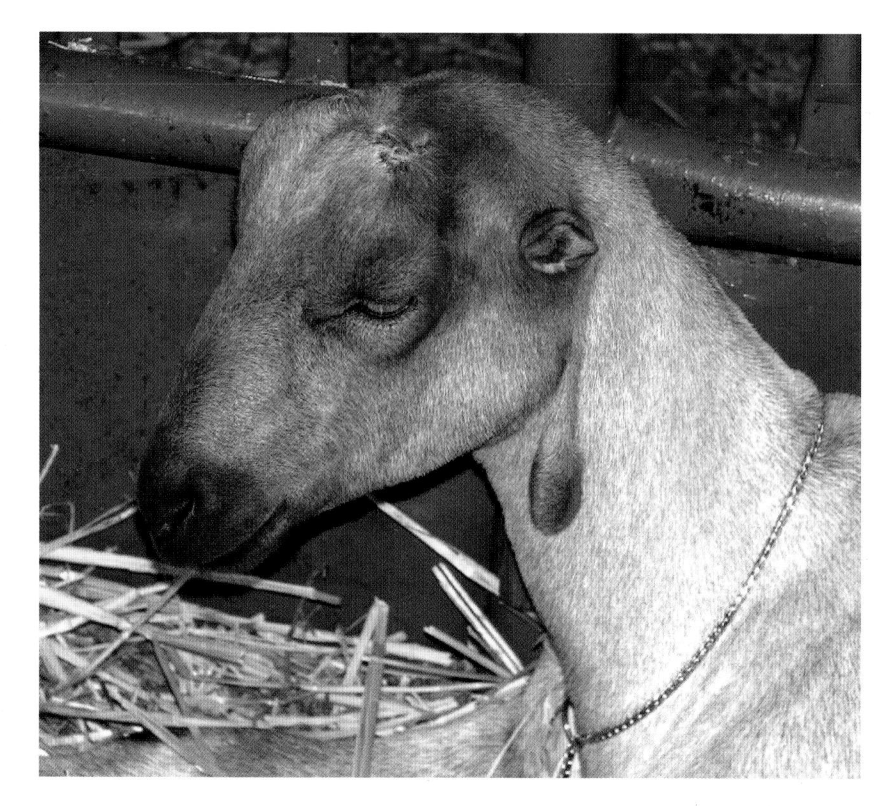

LaMancha's face is straight, the ears being its most distinctive characteristic. Any colour or combination of colours is acceptable with no preferences. The hair is short, fine and glossy. Roman noses, that are typically a characteristic of Nubian goats, are considered to be a moderate to serious breed defect in LaManchas. There are two types of LaMancha ear, the first being the gopher, in which the external ear is small and appears to be somewhat 'shrivelled'. There is no fold, but the external ear must always exceed 1in (2.5cm) in length. The second type is the elf ear, for which an approximate maximum length of 2in (5cm) is permitted. The end of the ear must be turned up or turned down, and cartilage shaping the small ear is allowed. LaManchas with a fold exceeding the permitted length cannot be registered as purebred animals.

Only males with gopher ears are eligible for registration; the reason for this is that two elf-eared animals bred together are likely to produce erect-eared kids, causing the uniqueness of the breed to be lost. In females, however, one type of ear has no particular advantage over the other.

MYOTONIC

Myotonic goats have many other names, including Fainting goats, Tennessee (Meat) goats, Nervous goats, Stiff-leg goats, Wooden-leg goats, and Tennessee scare goats. They are smaller and somewhat easier to maintain than larger breeds, which makes them desirable for raising on smaller farms. They are also kept as pets or show animals, being friendly and intelligent.

This is a goat whose muscles freeze, roughly for ten seconds, when it is startled. Caused by a hereditary genetic disorder, known as myotonia congenita, this generally results in the animal collapsing on its side, even though there is no pain. The goat does not 'faint', in any real sense of the word, as it never loses consciousness. Younger goats may stiffen and fall over, while older ones learn to spread their legs or lean against something for support, often continuing to run about with an awkward, stiff-legged shuffle. The myotonic condition appears to be strictly muscular and does not involve the nerves or the brain. The condition occurs in many species, including human beings, when they are unable to relax voluntary muscles after vigorous effort.

Slightly smaller than standard breeds, Myotonics weigh in from 60–170lbs (27–77kg), although some males can be as heavy as 200lbs (91kg). They have large, prominent eyes in high sockets. The hair can be short or long, with certain

OPPOSITE & LEFT: LaMancha's main characteristic is its almost non-existent ears.

individuals producing a good deal of body hair during colder months, although there appears to be no Angora strain. Common coat-colours are black and white, although most colours are found in the breed.

Classified as a meat rather than a dairy goat, the breed is listed as threatened by the American Livestock Breeds Conservancy; consequently, it is not used as often for meat as other meat breeds, in that its rarity makes the live goat more valuable. Even though some breed these animals as pets, fainting is a disorder that most producers try to keep out of their flocks' bloodlines, unless they are raising the goats specifically for this trait.

NIGERIAN DWARF

A miniature dairy goat of West African ancestry. Brought to the United States as food for large cats, such as lions, the survivors usually lived on in zoos. Nigerian Dwarfs are popular as hobby goats due to their easy maintenance and small stature.

RIGHT & OPPOSITE: Nigerian Dwarfs make good pets due to their amenability and small stature.

the milk excellent for cheese- and soap-making.

The goats are gentle and easily trainable, which makes them popular as pets. Many breeders sell bottle-fed babies (kids) that are bonded with humans, making them easy to manage. Nigerian Dwarfs can even be trained to walk on a leash and some enjoy coming into the house with their owners. Because of their compact size, they also make excellent 'visitor' animals for nursing homes and hospitals. Remember, however, that females or neutered males (wethers) make the best pets, as bucks can have an objectionable odour.

Nigerian Dwarfs produce a surprising quantity of milk for their size, ranging from 1–8lbs of milk per day, with an average of 2.5lbs (1.13kg). Since Nigerians breed year-round, it is easy to stagger freshenings (births), so that the entire herd is never dry. Thus, they are ideal milk goats for most families. Their milk has a higher butterfat content than that from full-sized dairy goats. It is usually about 5 per cent but can be as much as 10 per cent at the end of a lactation, a feature that makes

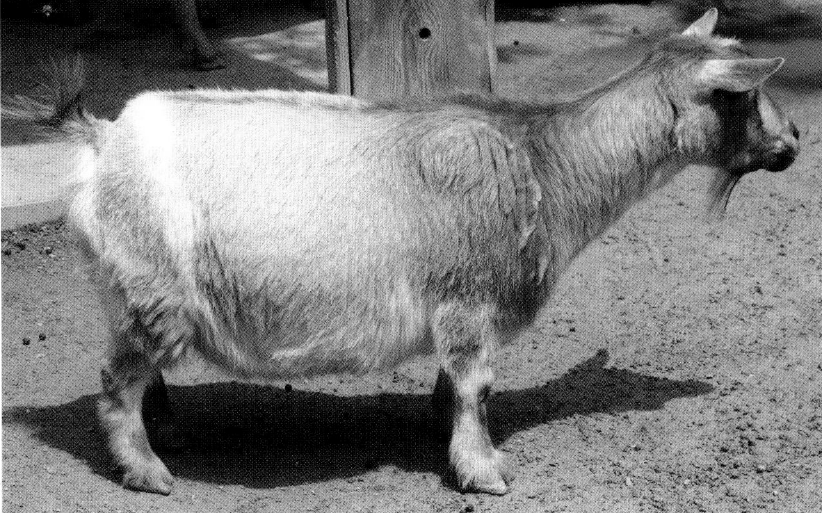

Nigerian Dwarfs come in many colours, including white, black, red, cream and patterns such as buckskin (brown with a black cape over the head and neck along with other black markings) and chamoisee (similar to an Oberhasli goat) with or without white spots. Some have white 'frosting' on the ears. Both the Nigerian Dwarf Goat Association and the American Goat Society websites include detailed colour descriptions, disqualifying features and conformation standards.

Although most are naturally horned, most breeders dehorn their goats at a early age (usually at less than 2 weeks) for the safety of the goat itself, its herd-mates, and its human carers. Blue eyes can occur, which is a dominant trait in goats.

NUBIAN

The Nubian or Anglo-Nubian originated in England as a cross between the Old English Milch Goat and Zariby and Nubian bucks imported from Egypt, India and Russia. They have been exported to most countries from England and in the United States are referred to simply as Nubians. Due to their heritage, Anglo-Nubians can survive in very hot climates and they have a longer breeding season than other goats. Considered a dairy or dual-purpose breed, Anglo-Nubians are known for the high butterfat content of their milk, although the breed produces a lesser volume of milk, on average, than other dairy breeds.

Anglo-Nubians are large, with males weighing in at 175lbs (79kg) and females coming in at 135lbs (61kg). Like most dairy goats, they are normally rendered hornless by disbudding within approximately two weeks of birth. The head is distinctive, the facial profile between the eyes and the muzzle being strongly convex (Roman-nosed). The ears are long (extending at least an inch beyond the muzzle when held flat along the face), wide and

OPPOSITE & BELOW: A Roman nose is desirable where the Nubian is concerned.

pendulous. They lie close to the head at the temple and flare slightly out and well forward at the rounded tips, forming 'bell' shapes. The ears are not thick, but the cartilage is well-defined. The hair is short, fine and glossy, with any colour or colours, solid or patterned, being acceptable.

While Nubians are often described as vocal animals, like human infants they are usually only making their needs known. They love human interaction and will even call for their owners. Nubians are also sometimes classified as stubborn, but they are simply highly intelligent animals which know what they like and dislike.

OBERHASLI

Formerly known as the Swiss Alpine, the name was changed in 1987 to coincide with the eponymous district of the Canton of Berne where the goat originated. The Oberhasli record for milk production is 4,655lbs (2111kg) in one year. It is a standardized colour breed, the goats having warm reddish-brown coats accented with a black dorsal stripe and black legs, belly and face. Occasionally a black Oberhasli appears as a result of a recessive gene. The American Dairy Goat Association (ADGA) considers black does acceptable, and will allow them to be registered, but black bucks are not permitted. The breed is about 2in (5cm) smaller in size than the other standard-sized (non-miniature) breeds. Temperament tends to be quiet and sweet-natured but alert, with vocalizing similar to the other Swiss-origin breeds.

First appearing as importations in the 1920s, the breed is still relatively rare in the United States, and until the early 1960s was still not accepted as a breed by the ADGA. Because of its rarity and close similarity in type to the

OPPOSITE: Nubian kids.

ABOVE RIGHT: The Oberhasli has a tendency to vocalize, like other Swiss-origin breeds.

multicoloured Alpine goat, Oberhasli registrations were lumped in with the American Alpines. A small number of them, however, were kept pure with their records intact, and the breed was kept alive, almost single-handedly, by Esther Oman, a breeder in California.

The hardy, thrifty Oberhasli is growing in popularity at shows (competitions based on selection of animals on the basis of body conformation) as their docile nature is appreciated, along with the fine colour of the coat, described as 'like the wood on the back of a violin'.

PYGMY

Originating in the Cameroon of Central and West Africa, Pygmy goats were imported into the United States from Europe in the 1950s for showing in American zoos as well as for research purposes. They were eventually acquired by private

breeders and quickly gained popularity as pets and exhibition animals due to their good nature and hardy constitution.

Hardy, alert, good-natured and gregarious, the Pygmy is a docile, responsive pet, a co-operative provider of milk, and an ecologically effective browser. It is an asset in a wide variety of settings, and can adapt to virtually all climates. Pygmy goats are often kept as pets in urban or suburban backyards, depending on the local regulation governing livestock ownership.

Pygmy goats are precocious breeders, bearing one to four young every nine to 12 months after a five-month gestation period. Females are usually bred for the first time at about 12–18 months, although they may conceive when they are only two months old if care is not taken to separate them early from the young males. Newborn kids will nurse almost immediately, begin eating grain and roughage within a week, and are weaned by three months of age. Being polyoestrous means that females cycle year-round, as opposed to other dairy breeds, which usually have problems maintaining a year-round supply; if milking is a priority, then a continuous supply can be obtained by breeding two females alternately.

Males weigh in at 40–80lbs (18–36kg), with females coming in at about 35–60lbs (16–27kg). Official

Pygmy goats are the most popular choice of pet, being good-natured, small and hardy.

coat-colours are caramel, grey agouti, brown agouti, black agouti and solid black. It is no longer necessary to assign a shade such as light, medium or dark. The colour formerly known as grey/brown agouti may be registered as grey agouti with other colours shown as random markings.

Pygmies are sociable animals and are happier in a herd atmosphere or with other goats as companions. Like all goats, Pygmies have stomachs with four compartments: the rumen, the reticulum, the omasum and the abomasum. As browsers, goats are similar to deer and enjoy variety in their diet, which includes woody shrubs.

PYGORA

A cross between Pygmy and Angora goats that produces three distinct kinds of fleece and is smaller in size than the Pygmy. The Pygora is a purposeful cross, developed by Katharine Jorgensen of Oregon City, Oregon. In 1987, the Pygora Breeders Association was formed in the United States, since when it has been registering and promoting the breed. Today, registered Pygora goats may not be more than 75 per cent registered Angora or 75 per cent registered Pygmy goat.

Pygoras live from 12–14 years and are commonly used as pets, along with being show-, breeding-, and fibre-producing animals. The fibre is used for spinning, knitting, crocheting, weaving, tapestries, and is also commonly used in clothing. Pygoras can also be milked, producing about 0.26 gallons (1 litre) of milk per day.

Registered Pygoras are used to produce a cashmere-like fleece

ABOVE RIGHT: Pygmy goats are good climbers, making secure fencing a necessity.

OPPOSITE: The Saanen is the largest of the dairy goat breeds.

(Classified as Type-C), a mohair-like fleece (Type-A), or a combination of the two (Type-B). Type-A fleece is composed of fibres averaging 6 or more inches in length that fall in ringlets. The fibre may occur as a single coat, but a silky guard hair is usually present. The fibres are typically fewer than 28 microns in diameter. Type-B fleece fibres average between 3–6in (150mm) in length with one, possibly two, guard hairs. The fibres are usually fewer than 24 microns in diameter. Type-C fleece is very fine, typically 1–3in (76mm) in length and fewer than 18.5 microns in diameter.

SAANEN & SABLE

Saanen goats are a white or cream-coloured, pink-skinned breed, named for the Saanen valley in Switzerland. Saanens are the largest of the dairy goat breeds, with the males weighing over 200lbs (90kg) and the females coming in at 150lbs (68kg) or more. The breed also produces the most milk (which will vary from one individual to another) that tends to have a lower butterfat content (about 2.5–3 per cent). The Saanen temperament is, as a rule, calm and

mild-mannered, to the extent that they have been described as 'living marshmallows', making them easier for children to handle and they are popular for showing. Saanens typically breed every year, producing one or two kids. Iif milk is required for personal use, it may be necessary to separate one kid from its mother if twins are born.

Since 1904 the Saanen has been one of the most popular breeds of dairy goat in America. Some are drawn by the pleasing aesthetics of a pasture full of uniformly white animals, while others are attracted by the Saanen's large size, vitality, herd compatibility and 'eager-to-please' temperament.

The Sable Saanen is not a crossbreed, but is a recessive expression of colour derived from the white Saanen. Sables can vary in colour from beige through to black, with almost any colour in between other that pure white. Sables are accepted as a breed in their own right in some breed clubs but not in others.

OPPOSITE & RIGHT: The Saanen is one of America's most popular dairy goats.

SAN CLEMENTE

A breed of domestic goat derived from feral goats isolated on San Clemente Island, which is located off the coast of southern California. The island is owned by the US government and is used and managed by the US Navy. Feral goats, probably of Spanish orgin, have inhabited the island for several centuries, possibly since the 1500s, while later introductions may have come from mainland Franciscan missions during the 1600 and 1700s, with farmers being responsible for later introductions.

The US Navy became responsible for the island in 1934 and, under a directive to preserve the endangered flora and fauna of the island, that were being threatened by the grazing of non-endemic species, sought the goats' removal. After initial trapping and hunting failed to eliminate them, the navy began a shooting programme of

extermination. This was blocked in court by the Fund for Animals, on the grounds that the goats were not threatening any endangered species, following which goats were adopted out on the mainland by the Clapp family and by the Fund for Animals. The US Navy was given the right to exterminate the remaining goats, and the last goat on San Clemente Island disappeared in April 1991.

San Clemente Island goats are small, fine-boned and deer-like, the males sporting outwardly-twisting Spanish-type horns. Although initial DNA studies indicated that they were not of Iberian origin, further studies show that they do indeed have Iberian bloodlines but have experienced a large degree of genetic drift. In that respect, they are very different from goats still existing in Spain and from the mainland Spanish goats of the USA.

LEFT: The San Clemente is a breed of domestic goat derived from feral goats isolated on San Clemente Island, located off the coast of southern California.

OPPOSITE: The ancient Toggenburg goat is an Alpine breed from Switzerland.

The breed is listed as a critically-endangered heritage breed by the American Livestock Breeds Conservancy. In 2008 the global population was approximately 400, living on the mainland USA and in western Canada.

TOGGENBURG

The oldest-known dairy breed of goats, the Toggenburg is named after the region in Switzerland where it originated. Toggenburgs perform better in cooler conditions. They are are medium in size, moderate in production, and have relatively low butterfat content (2–3 per cent) in their milk. They possess a general Swiss-marked pattern with various dilutions. The colour is solid, varying from light fawn to dark chocolate with no preference for any shade. Distinct white markings are as follows: white ears with dark spot in the middle; two white stripes down the face from above each eye to the muzzle; hindlegs white from hocks to hooves; forelegs white from knees downward with a dark line (band) below the knee acceptable; a white triangle on either side of the tail. Wattles, or small, rudimentary nubs of skin, located on either side of the neck, are often present in this breed.

The Toggenburg is the smallest of the dairy breeds, and is usually very friendly. A Toggenburg must have a compact body and a high, well-attached udder. It should have a straight or dished face, but never a Roman nose. Toggenburgs are generally a quiet and gentle breed, and they are good as pets.

The Toggenburg underwent a development programme when introduced to Britain, the result being that the British Toggenburg is heavier and has improved milk quality. By the middle of 2002, 4,146 Toggenburgs had been registered with the New Zealand Dairy Goat Breeders Association, representing over 8 per cent of registered dairy goats.

CHAPTER FOUR
SHEEP

BARBADOS BLACKBELLY

A breed of domestic sheep developed in the Caribbean. Although it is likely that the Barbados Blackbelly has African ancestry, there seems to be clear evidence that the breed, as seen today, was developed by the people on the islands from sheep brought by sea fairly early in the period after Europeans first arrived. The breed is raised primarily for meat.

In 1904, the US Department of Agriculture imported a small flock of sheep and transported them for study to Bethesda, Maryland. From that original flock, at least two distinct breeds have emerged in the US, leading to a great deal of confusion in the breed names. There are fewer than 200 purebred Barbados Blackbelly sheep in the US, in contrast with a large and growing population of a popular crossbreed, commonly referred to as a 'Barbado'. Whereas purebred Barbados Blackbelly rams and ewes are polled (hornless), the Barbado is most noted for the rams' regal rack of horns, and even some ewes may also have small horns. The horns appeared by crossbreeding Barbados Blackbelly with Mouflon and Rambouillet early after they were imported by the USDA. Rams with a large horn curl are commercially bred for use on private hunting ranches, where the attribute is prized by exotic game hunters.

Recognition of the name Barbado did not adequately define the characteristics that breeders sought, and the Barbados Blackbelly Sheep Association International

adopted a breed standard in 2004 and defined animals meeting this standard as 'American Blackbelly'. These striking, attractive sheep have become common in children's zoos and petting farms in the US. In addition to their exotic appearance, they are very easy to keep and breed.

Blackbelly sheep of both breeds are able to tolerate heat and have more stamina than most other sheep. They are fleet of foot and in many ways resemble deer. They are 'hair sheep', which means they do not grow wool but have coarse hair instead. If raised in cooler climates, they often develop a woollen undercoat that they shed in the spring.

Unlike most domestic sheep, Barbados Blackbellies will breed all year round. Because they are smaller and slower-growing than most woolled sheep, they are not a good choice for commercial production. However, there is a strong market for their lean and mild-flavoured meat, and they are popular with trainers of herding dogs. They are also disease-resistant and parasite-tolerant, and these genetic traits have created a demand for them in crossbreeding operations. They are the perfect homesteader's sheep because they do relatively well on poorer forage, can be raised with very little grain, and do not require intensive management. Blackbelly sheep range in colour from light tan to a dark mahogany red, with black stripes on the face and black legs, belly, inguinal region, chin and chest. Despite being goat-like in appearance, they are nevertheless true sheep.

OPPOSITE: The Barbados Blackbelly is a true sheep, despite its goat-like appearance.

BLACK WELSH MOUNTAIN

This is a colour type of the Welsh Mountain sheep. Occasionally, one will appear in a flock of other colours, but the Black Welsh Mountain is now often maintained as a separate strain. Like other Welsh Mountain sheep it is found mainly on the hills in Wales, but is also kept elsewhere.

Introduced into the United States in 1972, the fleece from the Black Welsh Mountain sheep has generated special interest among home spinners and weavers which has spread to other countries, such as Canada and Japan.

Apart from being wholly black, the Black Welsh Mountain is like other Welsh Mountain sheep in that it is small and hardy with no wool on the face or legs; the males are horned, but females are normally polled. It is a breed that is easy to maintain, having a natural resistance to disease. As a hill breed it is accustomed to grazing upland grass but will adapt

RIGHT: Occasionallly appearing in a flock of coloured sheep, the Black Welsh Mountain is a colour type of the Welsh Mountain sheep.

well to improved lowland management systems. Black Welsh Mountains are prolific and undemanding, hardy and self-reliant. They also produce premium-quality lean meat with an excellent meat-to-bone ratio and a full flavour.

BLUEFACED LEICESTER

Evolving from a breeding scheme by Robert Bakewell to develop the longwool sheep in the 1700s, the breed was originally known as the Dishly Leicester. It was developed over the next 200 years, when it became commonly known as the Hexham Leicester, due to the breed's early concentrations in Northumberland in the North of England. Known today as the Bluefaced Leicester, it is regularly crossed with many of the native British hill breeds, such as Swaledale, Blackface, Welsh Mountain and Cheviot, to produce the Mule ewe, which means any crossbred sired by a Bluefaced Leicester ram, and now makes up almost half of the UK's crossbred ewe population.

In the 1970s, Bluefaced Leicesters was exported to Canada, and exportation of frozen semen from the UK is now used to expand the breed's genetic diversity in Canada and the US. The breed is raised primarily for meat.

Bluefaced Leicesters have broad muzzles, good mouths and a tendency towards a Roman nose, also bright, alert eyes and long, erect ears. The colour of the head skin should be dark blue showing through white hair, although a little brown is generally acceptable.

OPPOSITE & BELOW: Bluefaced Leicesters are popular for their long, lustrous wool which is excellent for spinning and takes dye well.

Fully-grown rams weigh in at up to 240lbs (110kg) with ewes coming in at 200lbs (90kg). They have curly, threadlike wool which makes it considerably lighter than that of other breeds, and some fleeces weigh only 2.2–6.6lbs (1–3kg). The wool is classed as semi-lustre and fine, qualities which are usually passed even to crossbred offspring.

BORDER CHEVIOT

Also known as the South Country Cheviot, this is a breed of domesticated sheep native to the Cheviot Hills, bordering Scotland and England. Recognized as early as 1372, the breed is reported to have been developed from sheep that swam ashore from shipwrecked Spanish vessels that fled northwards after the defeat of the Armada. The breed is prized for its wool but bred primarily for meat.

The Cheviot is a distinctive white-faced sheep with a wool-free face and legs, pricked ears, black muzzle and black feet. It is a very alert and active sheep. Bred to look

after themselves, Cheviots need less husbandry than other breeds, and their ease of lambing and strong mothering instinct means fewer lambing problems. Their hard black feet make them less prone to foot rot, and they are relatively disease- and pest-resistant. They are a medium-sized breed, with males weighing in at about 180lbs (80kg) and the females coming in at around 88lbs (40kg).

Cheviot wool has a distinctive helical crimp, which gives it that highly desirable resilience. The fleece should be dense and firm with no kemp or coarse hairs. The rams can be horned.

BRECKNOCK HILL CHEVIOT

Also known as Brecon and Sennybridge Cheviots, this is a domesticated breed originating approximately 400 years ago in Wales, being the result of crosses between Welsh Mountain, Cheviot and Leicester breeds. The Brecknock Hill is primarily raised for meat. It was introduced into the US in 1838.

Brecknocks have erect ears with a white face and legs and a ruff of wool behind the ears. There is no wool on the face or legs below the knee or hock. Both sexes are polled, although rams occasionally come horned.

This is a medium-sized breed and more docile than the other Cheviot sheep, with the males weighing in at around 198lbs (90kg) and the females coming in at 132lbs (60kg). It comes in all sheep colours except spotted. The fleece weight is 3.3–5.5lbs (1.5–2.5kg) with some kemp. It is excellent for hand-spinning.

BRITISH MILK

Developed by Lawrence Alderson in Wiltshire and Northumberland, the British Milk sheep can now be found in the UK and Canada. The exact composition of the breed is debatable, but it is agreed that East Friesian, Bluefaced Leicester, Polled Dorset and Lleyn are at least present in its bloodline. The breed was released in 1980 and, as the name suggests, it is a prolific milk-producer. However, it is also known

OPPOSITE, RIGHT & PAGE 180: The wool of the Cheviot was once the base for the Border Tweed industry that paid the tenants' farm rents, but which has now declined. Today the wool is chiefly used in the carpet industry with a small amount being popular in the craft trade.

RIGHT: The California Red is a good all-purpose breed for small-scale producers.

for its high incidence of producing twins and triplets, and it is often raised for meat as well as for dairy purposes.

Varying from medium to large in size, British Milk sheep are large, white, polled animals that display good temperaments. Robust and active in nature, they are adaptable for all seasons and vary in weight from 176–265lbs (80–120kg).

British Milk sheep produce a heavy, lean carcase. Fleece weights range from 9lbs (4kg) in ewes and over 14lbs (6kg) in rams. The demi-lustre wool has a spinning count of 50–54s with a staple length of 5–7in (125–178mm). Ewes have a 300-day lactation period, producing 172–238 gallons (650–900 litres) of milk in the period, with 5.5–9.5 per cent fat content.

CALIFORNIA RED

A breed of domestic sheep developed in the United States in the 1970s. It is so named because the lambs are born all red, and they retain this colour on their faces and limbs into adulthood. Dr. David Spurlock, of Davis, California, crossed Tunis and Barbados Blackbelly sheep, and the California Red is subsequently a dual–purpose breed with many of the qualities of its forebears. It has the ability to lamb out of season, and is thus able to produce multiple lamb crops in a year. It also functions well in hot weather, and is polled in both sexes.

CANADIAN ARCOTT

A breed of domestic sheep native to Canada. The latter half of its name is an acronym for the Animal Research Centre in Ottawa, where it was developed along with the

Rideau Arcott. The Canadian Arcott is a synthesis of many breeds, but primarily of the Île de France and Suffolk. Like its progenitors, it is a terminal sire breed. Arcotts are a medium-frame, white-faced production breed largely kept for meat.

CHAROLLAIS

The breed is centred around Charolles in the Saône-et-Loire region of France, where it grazes alongside the famous white Charolais cattle. Since the time that it was

ABOVE: The Charollais is a French sheep, and can be seen grazing alongside Charolais cattle.

imported to Britain in 1976, it has grown in popularity and is currently probably the second most important terminal sire breed in the UK.

At maturity, rams weigh on average 300lbs (136kg), with ewes coming in at 200lbs (90kg). The breed has a fine fleece, a pinkish–grey face and comes naturally polled in both sexes.

CLUN FOREST

Similar in appearance to many of the British breeds of upland sheep, the Clun Forest is a multi-purpose animal, kept for meat, wool and milk. It is a medium-sized animal, known for its hardiness, longevity, fertility and good mothering abilities, the females usually producing twins which grow quickly due to the high butterfat content of their milk.

The most striking feature of the breed is the face, which is a rich, dark colour, rather narrow in width, and free from wool except for a top knot. The ears are held upright, giving the sheep an alert and lively appearance. Its powerful build is emphasized by its strong muscular neck and a long, broad back. Legs below the hock must be as free from wool as possible, while the fleece itself should be uniform in staple length, colour and texture from head to tail.

The breed takes its name from the old town of Clun and the surrounding forests in the south-west corner of Shropshire, England. One of the first mentions of the breed was in 1803 when the Rev. Joseph Plymley, writing on the agriculture of Shropshire for the Board of Agriculture, quoted from a previous report dealing with these forest sheep, remarking that the flocks upon the hills nearer to Wales were without horns and had white faces. In 1837 it was confirmed that the Clun Forest sheep were definitely white-faced and hornless, but that the breed was fast changing its appearance through crossings made with other local breeds, such as the Longmynd, Radnor & Shropshire, which was resulting in a darker colouring on the head.

In 1925 the Clun Forest Sheep Breeders Society was formed in Great Britain 'to secure the purity of lineage and fixity of type' and also to promote the virtues of the breed throughout the sheep industry. Gradually, flocks began to establish pedigree status and after the hiatus of the Second World War, the Clun entered its heyday, a period which extended from the mid 1940s to the 1970s. It is during this period that the Clun became the third most numerous purebred in Britain.

The Clun arrived in North America in 1970 when Tony Turner imported two rams and 39 ewes, to be followed by one further ram to Nova Scotia. In 1974 the North American Clun Forest Association was founded in Harrisburg, Pennsylvania. Following the original 1970 importation, Angus Rouse of Nova Scotia undertook two further importations in the late 1970s and mid 1980s. Since then US and Canadian borders have been closed to live sheep imports, but semen from some of the top Clun rams in the UK and the Netherlands have been imported in the mid 2000s.

In recent years the sheep has declined in numbers in its native Britain, but interest in the breed has grown steadily in North American and in the Netherlands. The breed is well-adapted to grassfed farming, and lambs consistently attain a market weight at seven to eight months of 100lbs (45kg) on grass alone. The meat is lean and of a mild flavour. Because Clun Forest ewes produce some of the highest butterfat in their milk of any sheep breed, Cluns are often crossed with dairy sheep, such as the East Friesian, to impart extra richness to cheeses, and to provide dairy flocks with larger marketable lambs.

Clun Forest fleeces are of moderate weight (6–8lbs/2.7–3.6kg) with a staple length of 4in (102mm) and are consistent from neck to britch. The fleeces are free from black or kemp fibres and hold an average spinning count of 58.

Columbias are one of the largest of the white-faced meat breeds. They require good feed to realize their full potential.

COLUMBIA

One of the first breeds of sheep to be developed in the United States. The product of the US Department of Agriculture and university research, it was intended to be an improved breed intended for the Western ranges of the country, where the majority of sheep-raising takes place. Beginning in 1912 in Laramie, Wyoming, Lincoln rams were crossed with Rambouillet ewes. In 1918, the foundation flock was moved to the US Sheep Experiment Station near Dubois, Idaho, for further refinement. Today's Columbia is a popular breed, with a heavy, white fleece and good growth characteristics. It is one of the larger breeds, and is often used for crossbreeding in commercial Western flocks.

Adult males weigh in at 225–300lbs (102–136kg), with females coming in at 150–225lbs (68–102kg). An average ewe fleece weighs 10–16lbs (4.5–7.3kg) with a yield of 45–55 per cent. The staple length of the wool ranges from 3.5–5in (90–127mm). The medium-range wool has a numeric count of 50s–60s and varies from 31–24 microns.

COOPWORTH

A breed developed by a team of scientists at Lincoln College (now Lincoln University) in Canterbury, New Zealand, to increase lambing percentages of Romney ewes when mated with Border Leicester rams. The breed comprises the second largest flock in New Zealand. Coopworths are also bred in Australia, parts of Europe and in the United States.

Coopworths are a dual-purpose, medium-sized longwool breed, with alert but quiet dispositions. The long face is usually clean, the head being bare or with a small top knot, and with a slightly Roman nose. Coopworths stand slightly taller than NZ Romneys and exhibit heavier muscling than the Border Leicester. The body is long with a good loin and hindquarters, light forequarters and a wide pelvis. Ewes lamb easily with very little outside assistance required.

The fleece, with pointed locks, has a well-defined crimp with a bright lustre, a spinning count of 44–48 (35–39 microns), and a staple length of 5–8in (127–203mm). While only white Coopworths may be registered in New Zealand and Australia, both white and natural-coloured sheep are accepted for registration in the United States and Canada. It is important to note that it is not unusual to see differences in appearance between individual animals, because selection based on measured performance, rather than phenotype, has traditionally been the basis for registration. For this reason, several wool styles are considered acceptable. These would be similar to Border Leicester and NZ Romney wool types.

CORMO

An Australian breed developed in Tasmania by crossing Corriedale rams with superfine Saxon Merino ewes in the early 1960s. The name Cormo is derived from the names of two of the parent breeds, Corriedale and Merino. The breed was fixed through intense selection criteria, assessed by objective measurement. Cormo sheep have polled, open faces, are possessed of a fast-growing, medium frame carrying a fleece of about 18–23 microns in diameter. High fertility is an additional attribute.

The breed is mostly found in the south-eastern states of Australia, but Cormos have also been exported to Argentina, China, the USA, Italy and Belgium.

CORRIEDALE

The dual-purpose Corriedale is the oldest of all the crossbred breeds, being a Merino-Lincoln cross developed almost simultaneously in Australia and New Zealand. The breed was developed between 1868 and 1910 and was gradually distributed to many of the sheep-raising areas of the world, having been first brought to the United States in 1914. The Corriedale was later used as one of the parents of the US-developed Targhee breed.

The goal was to develop a breed that would thrive in lower rainfall areas and supply long staple wool.

The Corridale, a Lincoln-Merino cross, was developed almost simultaneously in Australia and New Zealand from around 1868.

James Little was the original breeder and the name comes from a property in New Zealand's South Island, where he conducted his work, with encouragement from New Zealand and Australian Land Company superindent, William S. Davidson.

Corriedales have a long lifespan and are hardy and evenly-balanced over the whole of the body. A medium to large sheep, it is especially suited to all types of grazing in lighter rainfall areas; it is widely used in New Zealand on every sort of terrain, from intensively grazed lowlands and plains to all but the very highest mountain locations, and in similar environments worldwide. Ewes have high fertility and make docile, easy-to-maintain mothers.

Corriedales produce big fleeces and top-quality meat. The wool is bright, dense, bulky and soft-handling, ranging from 31.5–24.5 microns. Fleeces from mature ewes weigh 10–17lbs (4.5–8kg) with a staple length of 3.5–6in (90–152mm). After cleaning, a yield of 50 to 60 per cent of the raw fleece weight is common. Mature males weigh in at 175–275lbs (79–125kg), with females coming in at 130–180lbs (59–82kg).

COTSWOLD

A dual-purpose breed of domestic sheep, originating in the Cotswold Hills of the southern English Midlands. As of 2009, this longwool breed is relatively rare, and is categorized as a 'minority' breed by the UK's Rare Breeds Survival Trust.

Some claim that the Cotswold breed was already in the region when

RIGHT, OPPOSITE & PAGE 190: The Cotswold sheep originated in the Cotswold Hills in south-central England.

the Romans arrived circa 54 BC. In July 1964 a Roman-sculpted replica of a sheep's head was reported as having been unearthed near Bibury Church in Gloucestershire; the resemblance to a modern Cotswold sheep is striking. Legend tells how there used to be sheepcotes, or solidly-built folds, and that these gave their name to the hills, i.e., 'Cotteswealds', in Anglo-Saxon, meaning exposed, windswept hills (wealds or wolds) dotted with cotes. Thus the Cotswold Hills got their name from the flocks, which in turn got their name from the hills.

From the late 1500s to the early 1600s, Cotswolds were commonly noted as having slightly golden-coloured wool, dark colours being exceedingly rare. This trait gave them

the nickname 'Golden Fleece Breed'. According to W.S. Harmer of the British Cotswold Sheep Society, writing in 1892, the Cotswold is the only breed to have been associated with the fabulous cloth of gold of antiquity, which is said to have prompted Florentine merchants, at least as far back as the 13th century, to travel to England to buy large quantities of the shiny, linen-like wool. Cotswold wool was used as a substitute for linen, woven with exceedingly fine wires of real gold to make special garments for ancient priests and kings, described in the Book of Exodus.

The breed was first introduced into the United States in 1832, by Christopher Dunn of Albany, New York. (Importation records of

Cotswolds only date back to that era.) While Christopher Dunn imported only one Cotswold ram to mate with his English Leicester ewes, the resulting crosses were so impressive that they prompted William Henry Sotham (bankrolled by the Hon. Erastus Corning, also of Albany) to make extensive importations of the sheep from the flock of William Hewer of Northleach, Gloucestershire, England. Another early contributor to American flocks was Charles Barton, of Fyfield, North Leach, England, who had family records of Cotswold pedigrees going back to at least 1640.

As with other longwool breeds, the Cotswold was often used for crossbreeding in early times. By 1914 over 760,000 had been recorded in the US and Canada by the American Cotswold Record Association. The breed was seen as a way of adding staple length to other breeds while not compromising the size of the carcase or thickness of the wool. The main reason for the Cotswold's early popularity over other lustre longwools in the US was because it did not require 'high feeding', in other words, large amounts of grain, in order to make good growth. It is claimed that the largest recorded representative of the Cotswold breed in America was Broadfield's Pride, owned by Charles Mattocks during the 1870s. This sheep was born in 1870 on the farm of William Lane of Gloucestershire, England, and attained the enormous weight of 445lbs (202kg); several of the yearling lambs sired by him attained weights of 280–300lbs (127–136kg).

In 1989, the Black Cotswold, a separate breed, was recognized and a new breed association was formed to assist growers in propagating the breed. The Black Cotswold can in fact be any colour, including white, as long as it is related to black sheep. In over 130 years of registering Cotswold sheep, no sheep registered with the American Cotswold Record Association has descended from coloured ancestors. Some old-time 'black' Cotswolds historically refer, in one form or another, to crosses such as those originally noted in the William Large flock of the early 1800s in England. Those sheep were the product of extensive crossings with English Leicesters, a breed more often known for possessing coloured wool.

Cotswolds do not have the flocking instinct, preferring instead to spread out and graze pastures more uniformly. Some strains of the breed are not as prone to internal parasites as others, provided their grazing forages are not excessively short. Cotswolds are usually calm and friendly. They mostly have white faces, which are occasionally mottled with light grey or tan hairs. Small black spots are permissible on the 'points' (the non-woolly portions of legs, ears, and face). Coarse hairs or kemp must be absent from the wool, which must itself be white. Cotswold hooves should be black, but are sometimes streaked with undesirable light or translucent colour. Foot rot is very uncommon in the breed. Rams occasionally have small scurs (highly discouraged) but no Cotswold should ever have full horns.

Cotswold lambs are very hardy, and provided that the ewes are not overfed during gestation, the lambs will have small heads at birth, making them more easily born than some other breeds. Cotswold ewes usually have a 'narrow flank', once thought to be a weakness until it was observed that it actually assists in directing unborn lambs towards a 'normal' birth presentation. Most Cotswold ewes produce abundant milk, which can be a problem if they only have a single lamb. Pulpy kidney disease causes convulsions and sudden death, mainly in well-fed lambs or young sheep. If a ewe is vaccinated before lambing, the lamb will be protected for up to eight weeks as it receives antibodies in the milk. The disease can also occur when sheep are fed on highly nutritious pasture or their grain is suddenly increased, causing a rapid build-up of toxins in the intestine.

Cotswolds live for a respectably long time, and it is not uncommon for ewes to have twins each year until well after they are ten years old. The Cotswold is today considered to be a fairly slow-growing sheep, because too much grain often kills it. However, it is quite unique in its ability to thrive where other lustre longwools might starve to death.

The breed has a very mild-flavoured meat, and the mutton of Cotswold sheep tends to be far less gamey than even the young lamb of many other breeds, and it is enjoyed even by those professing to hate eating lamb. (Lamb in general is considered to be the most tender, most hypoallergenic of the common meats, as well as being exceedingly easy to digest. This makes it ideally suited to small children, the elderly and athletes in training.)

Today, Cotswold wool is especially luxurious when hand-combed to make a true worsted roving. In the genuine article there is little or no 'itch', because all the tips of the fibres (as they grew on the sheep) point in one direction and the end sheared from the sheep points in the other direction. This produces a knit very like mohair; in fact, Cotswold wool has often been called 'poor man's mohair'. The wool can put on 8–12in (20–30cm) of growth in a year, and if not shorn promptly in early spring may become matted. It has a Bradford (spinning) count of 36s–44s, most commonly around the 40s. Generally, the tighter the curl of the fleece, the slenderer the wool. Because the wool is so long and parts on the sheep's spine, cold rains are liable to cause health problems; low temperatures and heavy snows are not problematic, however, where the sheep are concerned. Where the wool parts along the backbone, it is possible for flies to bite the skin, and warble flies have been known to cause sterility in rams when they occur near to the sheep's head.

DEBOUILLET

A breed of domestic sheep originating in the US state of New Mexico. It was developed in the 1920s through crossings of Rambouillet and Delaine Merino sheep, the breed's name being a portmanteau word describing the two ancestors. The breed is primarily raised for its wool.

Specifically adapted to the arid ranges of the American South-West, the Debouillet is a medium-sized sheep with long, fine wool. The fleece from mature ewes weighs from 10–18lbs (4.5–8.2kg) with a 35–50 per cent yield. The stable length of the fleece is from 3–5in (76–127mm) with a numerical count of 62–80, which is 18.5–23.5 microns.

Ewes are polled and rams may or may not have horns. Mature rams weigh in at 175–250lbs (79–113kg), while ewes come in at 125–160lbs (57–73 kg).

The Dorper has the characteristic black head and neck and the White Dorper is white-headed.

DORPER

A South African breed developed by crossing Dorset Horn and Blackhead Persian sheep in the 1930s, the name 'Dorper' being a coupling of the first syllables of the names of the parent breeds. The breed was created through the efforts of the South African Department of Agriculture to produce a meat sheep suitable for raising in the more arid regions of the country. It is now farmed in other

areas as well, and is the second most common sheep breed in South Africa. Other breeds, such as the Van Rooy, are also believed to have contributed to the development of the breed. The Dorper Sheep Breeders Society of South Africa was founded in 1950.

The Dorper received its hardiness, thriftiness and adaptability from the Blackhead Persian sheep, a fat-tailed desert breed from Arabia.

The Dorper is an easy-to-maintain sheep that produces a short, light coat of wool and hair that is shed in late spring and summer. It has high fertility and good maternal instincts, combined with a high growth-rate and hardiness. It has the characteristic black head, while the White Dorper has a white head.

The Dorper adapts well to a variety of climatic and grazing conditions and has spread from the arid areas of its native land to all parts of the republic. It is reputed to do well in various range and feeding conditions and is also suited to intensive feeding. The breed is said to be able to graze and browse, which suggests it will consume plants seldom eaten by Merino sheep.

In Australia, Dorpers are now farmed throughout the arid and tropical areas as well as in the high-rainfall southern states, thriving even in the extreme cold and wet of Tasmania. Dorpers can be run as a replacement or (with suitable management) as a complementary flock to Merinos, particularly as shearing costs continue to rise and wool prices fall.

DORSET

It is said that Merino sheep came to south-west England with the Spanish Armada in the 16th century, where they were crossed with the Horned Sheep of Wales, which produced a desirable all-purpose animal which met the needs of the time. Thus began a breed which spread over Dorset, Somerset, Devon, and most of Wales and which were known as Horned Dorsets. In the USA they are called Dorsets.

The Cornell University Sheep Program developed and teaches the STAR system to promote frequent lambing with Dorsets, maintaining a research and teaching flock five miles south of Dryden, New York. The Dorset has a white face with close short fleece. It has a solid build, with broad back and short legs. Originally, both rams and ewes were horned. The polled Dorset originated in a herd at North Carolina State College, in Raleigh, North Carolina, and a registry of it was established in 1956. Since then, breeders of the polled variety have outnumbered breeders of Horned Dorsets.

Dorsets, whether polled or horned, are all-white sheep of a medium size, with the good body length and muscle conformation to produce a desirable carcase. The fleece is very white, strong, close and free from dark fibre. Fleeces average 5–9lbs (2.3–4kg) in the ewes, with a yield between 50 and 70 per cent. The staple length ranges from 2.5–4in (63.5–102mm), with a numeric count of 46s–58s. The fibre diameter ranges from 33–27 microns. At maturity, rams weigh in at 225–275lbs (102–125kg), with ewes coming in at 150–200lbs (68–90kg), although some in show condition may very well exceed these weights. Dorsets are one of the few breeds that carry the 'out-of-season' breeding characteristic. The ewes are good mothers, good milkers and multiple births are not uncommon. Dorsets work well in commercial situations, not only in the ewe flock but also in the role of terminal sires.

The Somerset is a related breed, but is larger and has a pink nose. The Portland is a smaller, primitive Dorset breed that takes its name from the Isle of Portland and was once common all over Dorset.

EAST FRIESIAN

The German Friesian, or East Friesian, is the best-known and most important of the Friesian sheep breeds, and is found in small numbers in many parts of the country as a producer of household milk.

The Friesian breeds originated in the ancient region of Friesland, extending along the North Sea coast westward from the Weser river in the north-east of Germany, along the northern coast of the Netherlands and south to the Schelde (Scheldt) river at the border of the Netherlands and Belgium. Offshore is a fringe of islands including the West Friesians, belonging to the Netherlands, the East Friesians belonging to Germany, and the North Friesians that are divided between Germany and Denmark.

The East Friesian sheep, along with Friesian cattle, including the Holstein, produce possibly the highest milk yields of any livestock.

The Friesian sheep breeds are of the marsh-type, and include the East Friesian Milk Sheep (Deutsches Friesisches Milchschaf), the Dutch Friesian Milk Sheep (Fries Melkschaap) from West Friesland, and the Zeeland Milk Sheep (Zeeuwes Melkschaap) from the Zeeland province. These breeds are similar in appearance, being polled in both sexes, with white wool and white faces, ears, and legs all devoid of wool. Their most distinctive physical feature are their 'rat-tails', which are thin and also devoid of wool. Prints from the early 19th century depict a 'short-tailed' Friesian sheep which suggests a link with the Northern European short-tail breeds, the Finnsheep or Finnish Landrace and the Romanov, which is consistent with the high prolificacy of the Friesians, which often give birth to three and four lambs at a time.

The East Friesian is possibly the most milk-productive sheep in the world, yielding an average of 79–159 gallons (300–600 litres) per 200–300-day lactation, although there are reports of yields reaching more than 900 litres, this being highly dependent on genotype and feeding intensity. It is also a high-fertility breed, used to increase flock fertility and for breeding milking ewes.

The East Friesian's white wool is mainly used for carpet-making, the fibre diameter being approximately 35–37 microns. The staple length is 4.7–6.3in (119–160mm) and fleeces range from 8.8–11lbs (4–5kg).

FINNSHEEP

Also known as the Finnish Landrace, the Finnsheep is a breed with a high incidence of multiple births, it being common for a ewe to have three, four, or even five lambs at a time. In North America there have been several instances of births of seven lambs, and the record in Finland is nine live births. The lambs, though often small, are vigorous at birth and grow well,

maturing early and can be mated at six months. Ewes commonly breed out of season and some may lamb twice in a year. The Finnsheep belongs to the group of Northern European short-tails, which also includes the Friesian, Shetland, Icelandic, Romanov, Spelsau, Swedish Landrace and several other breeds.

The Finnsheep is often used in crossbreeding programmes to increase lambing percentage, and Finnsheep bloodlines are found in many of the newer breeds. In the USA the breed is promoted by the US Finnsheep Breeders Association. Finnsheep were first imported to North America by the University of Manitoba, Canada, in 1966

Finnsheep have a range of fleece colours similar to those of Shetland and Icelandic sheep, with white being genetically dominant and the most commonly occurring. Black and black/white piebald (spotted) sheep are also fairly common, while browns, greys and fawns are scarce in the US at this time. White stockings and tail-tips, white crowns or facial markings, including the panda-like eyespot pattern, are common in coloured Finnsheep.

Finnsheep ewes are well-known for multiple births, the record being nine at one time.

While there is a range of wool fineness across individual Finnsheep, the American Sheep Industry's American Wool Council ranks Finnsheep in the fine end of the medium-wool category. The wool has a soft handle, a moderate crimp and a high lustre.

Australian Finnsheep are universally white, the wool having superior length, softness, better radius of curvature and reduced itch factor. In Australia, wool quality and length have improved to such an extent that there are now sheep that can be shorn twice a year and whose advantageous wool characteristics have been extensively incorporated into the Merino flock.

GULF COAST NATIVE

Believed to be descended from flocks brought to the New World by the Spanish conquistadors in the 1500s. The genetic origins of the Gulf Coast breed are not known, since a variety of types and breeds of sheep existed in Spain at the time. But Churra sheep, multi-purpose animals used for meat, milk, and coarse wool, were commonly brought to the Americas by the Spanish, and may well have contributed to the breed's foundation. At the same time, the Gulf Coast's fine wool suggests a contribution from pre-Merino types.

The American Sheep Industry Association considers the Gulf Coast Native to be one of the oldest breeds in North America. Little is known concerning the breed before the 19th century, although it is known to have existed for centuries. As late as 1717, 2,500 Spanish sheep were brought from Mexico City to Los Adaes near Natchitoches, Louisiana, and importations of French sheep and possibly other breeds may have mated with the Spanish sheep.

Descendants of these sheep developed largely through natural selection under humid semitropical range conditions in the Gulf Coast areas of East Texas, Louisiana, Mississippi, Alabama, Georgia and Florida. Prior to the Second World War, hundreds of thousands were allowed to range freely in unimproved pastures and sugar cane fields. Twice a year they were rounded up for shearing and to mark the lambs. After the war, the emphasis on high-input agriculture caused the sheep industry to turn to breeds which were larger in size and produced more wool and meat. This made the numbers of Gulf Coast sheep decline dramatically, endangering their very existence. Now, with renewed interest in low-input sustainable agriculture, the popularity of Gulf Coast sheep is reviving. The breed has been known by various names, such as Florida Native, Louisiana Native, Common Sheep, Woods Sheep, Native Sheep and Pineywoods Sheep. Remnants of these sheep survive today and are known as Gulf Coast Natives.

HAMPSHIRE

Also known as the Hampshire Down, the breed was developed, around 1829 from crossings of Southdown with the Old Hampshire, Old

OPPOSITE: Gulf Coast sheep have valuable performance characteristics that suit them to low-input production, such as resistance to gut parasites, foot rot and other diseases.

ABOVE: A flock of Hampshire sheep.

Wiltshire and the Berkshire Nott breeds native to the open, untilled, hilly stretch of land known as the Hampshire Downs.

John Twynam, a Hampshire farmer, crossed his then Hampshire flock with Cotswold rams in around 1829, the resulting half-bred rams being compact and stocky animals. From around 1835 they were sold into six or more of what were to become the first recognized pedigree

Hampshire Down flocks in the United Kingdom. In 1889 the Hampshire Down Sheep Breeder's Association was established in Salisbury, England, where it continues to be active.

That same year (1889) the American Hampshire Down Sheep Association was also organized, now known as the American Hampshire Sheep Association. Hampshire sheep had been reported in the US in around 1840 although there are no records to indicate that any survived the Civil War. In around 1865–70 Hampshires were again imported from England but the first authentic record of importations was made in 1879. For over a century, selected Hampshires have been bred to fill specific US industrial needs. Hampshire sheep have the genetic ability to convert forage efficiently into meat and fibre and are adaptable and productive in various geographic regions of the United States.

The South Downs of Hampshire and Sussex had long had sheep which had dark-brown or black legs, matured early, produced the best of mutton and a fine quality of medium wool. The original Hampshire was larger, coarser, but hardier, being slower to mature, with inferior flesh, and a longer but coarser wool. The Southdown had always been remarkable for its power of transmitting its special characteristics to its progeny through other kinds of sheep, and hence it soon impressed its own characteristics on the Hampshire. The horns of the original breed have disappeared; the face and legs have become dark, the frame has become more compact, the bones smaller, the back broader and straighter, the legs shorter, and the flesh and wool of better quality, while the superior hardiness and greater size, as well as the large head and Roman nose of the old breed, still remain. Hampshires of the 1890s matured early and fattened readily. They clipped from 6–7lbs (2.7–3.2kg)of wool, suitable for combing, which was longer than Southdown wool even though it was less fine.

The resultant mutton had a desirable proportion of fat and lean, and was juicy and fine-flavoured; the lambs were large and were usually dropped early and fed for market. Indeed, the Hampshire may be considered to be a larger and a trifle coarser and hardier Southdown. The breed was occasionally crossed with Cotswold, when it produced a wool more valuable for worsted manufacturers than the pure Cotswold itself. There is little doubt that, in addition to Southdown, the Hampshire has a dash of Cotswold blood in its composition.

The Hog Island was developed from feral breeds. It is extremely rare and desperately needs more breeders to be involved in its conservation.

HOG ISLAND

A breed developed from feral animals on Virginia's Hog Island beginning in the 17th century. In 1933, a hurricane destroyed most of Hog Island, and with the inhabitants abandoning the settled areas, many sheep were left to fend for themselves and reverted to a feral state. In the 1970s, the island was bought by Nature Conservancy, and most of the sheep were removed in order to prevent overgrazing; the few that remained died out or were eventually removed in a second round-up. Today, the breed is extremely rare, and with fewer than 200 registered animals it is listed as

'critical' by the American Livestock Breeds Conservancy. Both ewes and rams may be horned, and they are generally white with spotted faces, though black individuals appear from time to time. They are fine-boned, and weigh around 100lbs (45kg).

ICELANDIC

Descended from the same stock as the Norwegian Spelsau, and brought to Iceland by the Vikings, Icelandic sheep have been bred for a thousand years in a very harsh environment. Consequently, they are efficient foragers. The Icelandic is one of the Northern European short-tail breeds that exhibit flattish, naturally short tails. It is a medium-sized breed, generally short-legged and stocky, with face and legs devoid of wool. It is not a docile breed, and it is generally alert and fast on its feet. Icelandics exist in both horned and polled strains. Generally left unshorn for the winter, the breed is very resistant to cold weather. Multiple births are very common, with a lambing percentage of 175–220 per cent. There also exists in the breed the Thoka gene, named after the ewe first exhibiting the trait, and those carrying it have been known to give birth to triplets, quadruplets, quintuplets, and even sextuplets on occasion.

Ewes can be mated at five to seven months, although many farmers wait until the ewe's second winter before allowing it to breed. They are seasonal breeders and come into oestrus in around October, with a breeding season lasting up to four months. Rams mature early and can begin breeding as early as five months.

Throughout the world, the breed is famous for its wool, but in Iceland it is bred almost exclusively for meat. The fleece has an inner and outer layer typical of the more primitive breeds, the fine undercoat being the *thel*, and the long, coarser outercoat the *tog*. The average fleece weighs 4–5lbs (2–2.4kg) in grease. Due to the length of fibre, the openness of the wool, the natural colours and the versatility, fleeces are usually sold to hand spinners. The *thel* is down-like, springy, lustrous and soft. The longer *tog* coat is similar to mohair, being wavy or corkscrew rather than crimped and is good for worsted spinning. The natural colours vary from snow white through several shades of grey to pitch black, as well as several shades of moorit (brown) to brownish-black. Some individuals will also show mouflon, badgerface patterns with several combinations of colour and patterns. Bi-coloured individuals are also fairly common.

OPPOSITE & LEFT: The Icelandic's extreme hardiness allows it to survive its inhospitable environment.

ÎLE DE FRANCE

A breed native to the Île-de-France, a region of north-central France incorporating Paris. It was first developed at a French veterinary college in the 1830s through crosses of Dishley Leicester and Rambouillet, and was originally known as the Dishley Merino. A breed association was formed in 1933, and the sheep was rigorously tested early in its breeding for meat characteristics and maternal qualities.

Today the Île-de-France is one of the top meat breeds worldwide, and is present in South Africa and the Americas as well as in Europe. The male is primarily used as a terminal sire, but the breed is also occasionally found in dairy herds in the United States. It is a large, naturally polled breed with a white fleece.

JACOB

An English breed with a possible Scandinavian background. Only relatively recently has the older name, the Piebald, been superseded by the name Jacob. The sheep are believed to be the descendants of those brought to the British Isles by the Norsemen, who travelled the seas with livestock

as they sought new lands to colonize and farm, leaving their domestic animals behind.

It is a well-loved myth that Jacobs are related to the fat-tailed sheep of the Middle East, but there are neither genetic markers nor is there any historical evidence to bear this out. The colourful stories have been perpetuated mostly by the Biblical story of Jacob, which tells how he took every speckled and spotted sheep as his own from Laban's flock. Jacobs were formerly known as 'Spanish' sheep, but like so many British breeds with an unclear history, the legend persists that the Jacob washed up in England after the wreck of the Spanish Armada in the 16th century. However, there are records of spotted sheep (including four-horned sheep) existing in Britain at about the same time that make it clear that the breed is indeed very old.

A breed of unimproved multi-horned sheep, patterned with black and white spots, Jacobs are grown for their wool, meat and hides, also to be kept as pets.

Jacobs were probably first imported to Canada and the United States as zoo animals, due to their exotic appearance. Some individuals acquired the sheep from zoos around the 1960s, but the breed remained very

OPPOSITE & ABOVE: Jacobs are one of the few breeds that can have horns numbering from two to four.

rare in North America for several decades. Identification of the breed as endangered, and their ensuing registration, began in 1985. As of 2009, Jacobs are listed as 'threatened' by the American Livestock Breeds Conservancy, which means that it has fewer than 1,000 annual registrations and has an estimated fewer than 5,000 global population. The Rare Breeds Survival Trust in the UK do not view Jacobs as being at risk, since there are in excess of 3,000 registered breeding females in Britain. Jacobs in Britain were often kept as decorative animals, grazed in parks by the landed gentry, which probably contributed to their survival.

Jacobs are shorn once a year, although some individuals will show a natural 'break' in the fleece in spring, which sometimes leads to a natural shedding of the fleece, particularly around the neck and shoulders. The wool is valued by hand spinners, provided that it is free from kemp.

KARAKUL

Possibly the oldest breed of domesticated sheep, archeological evidence indicates the existence of the Persian lambskin as early as 1400 BC and carvings of a distinctly Karakul type have been found on ancient Babylonian temples. Although known as the 'fur' sheep, the Karakul provided not only the beautifully patterned and silky astrakhans of the young lambs, but they were also a source of milk, meat, tallow and wool, a strong fibre that was felted into fabric or woven into carpets.

The breed is named after the village of Karakul which lies in the valley of the Amu Darja river in the former emirate of Bokhara, West Turkestan. The region is one of high altitude with scant desert vegetation and a limited water supply, which gave the breed a hardiness and ability to thrive under adverse conditions, which is feature of the Karakul sheep to this day. Karakuls are also raised in large

LEFT: The Karakul may possibly be the oldest breed of domestic sheep dating back to 1400 BC.

OPPOSITE: The Katahdin is an American breed developed in Maine.

numbers in Namibia, having been first brought to the country by German colonists in the early 20th century.

Very young, or even foetal Karakul lambs, are prized for their pelts, which go under various names, including *karakul* (also *caracul*), *swakara* (South West African), *astrakhan* (Russian), Persian lamb or *agnello di persia* (Italian), *krimmer* (Russian) and *garaköli bagana* (Turkmen). The newborn lambs have a tight, curly pattern of hair, and must be under three-days-old when they are killed, or they will lose the black colour and soft, tightly-curled coils of fur. Dark colours are dominant and lambs often darken in colour as they age. Foetal Karakul lambs' pelts are called broadtail, *breitschwanz* (German) and *karakulcha*, and they are used (rightly or wrongly) to create various clothing items in haute couture and elsewhere.

KATAHDIN

The Katahdin is a breed of domestic sheep developed in Maine, USA, mostly in the second half of the 20th century. The breed was developed by

Michael Piel who, after reading an article in the February 1956 *National Geographic* magazine, imported selected St. Croix sheep chosen by Dr. Richard Marshall Bond and crossed them with various other breeds, selecting lambs based on hair coat, meat-type conformation, high fertility and flocking instinct. The Katahdin sheds its winter coat, therefore it does not have to be sheared. Its popularity in the US has increased in recent years due to low wool prices and high shearing costs.

LACAUNE

The breed originated in the Aveyron and Tarn departments and the surrounding regions of France, located to the south of the Massif Central. The Lacaune is the most numerous sheep breed in France, and is especially important to French agriculture, being the sole breed whose milk is preferred for the production of the famous and expensive Roquefort cheese.

The Lacaune represent an excellent and robust dual-purpose breed, characterized by good milking performance with high butterfat and milk protein, together with high daily gains in lambs, producing good carcase quality, with mild-tasting, good-coloured meat.

ABOVE: The Lacaune is a French breed whose milk is important for making Roquefort cheese.

OPPOSITE: The Leicester Longwool: the US founding fathers, George Washington and Thomas Jefferson, took pains to import Leicester rams from England to improve their stock.

LEICESTER LONGWOOL

A large-framed, dual-purpose sheep carrying a heavy, long-stapled fleece, the Leicester Longwool is a sturdy, efficient and adaptable breed. It thrives in a variety of climatic conditions, making good use of marginal forages. Rams average 250lbs (113kg), and ewes 180lbs (82kg). Ewes are good mothers, giving plenty of rich milk, with a lambing percentage of 120–150 per cent and higher in selected flocks.

America's founding fathers, George Washington and Thomas Jefferson, both had large flocks of sheep and took pains to bring in good Leicester rams from England to improve their stock. In fact, Leicesters are world travellers, having been exported to mainland Europe, North and South America, Australia and New Zealand. In 1826 the Leicester Longwool was one of the first pure breeds to be imported into Australia. Today it is known there as the 'English Leicester' and it is valued as a crossing sire to improve the carcase qualities of fine-wool breeds.

During the late 19th and early 20th century the Leicester gradually fell out of favour as it was replaced by newer breeds. By the 1930s it was nearly extinct in North America. Today it is classified as 'rare' by the American Livestock Breeds Conservancy, but it is enjoying a revival of interest due to the quality of its lustrous fleece and other desirable features.

LINCOLN

The Lincoln, sometimes called the Lincoln Longwool, is an English breed. The present-day Lincoln is said to be the result of crossings between Leicesters and the coarse native sheep of Lincolnshire. Not all breeders cared for the greater refinement and increased quality that Leicester blood introduced, but in the end, market demand resulted in improved carcases and higher quality wool.

The old Lincolnshire sheep was gradually modified by crossbreeding and selection towards a more useful sheep than the extremely large and thin-fleshed animal originally found in the area. Many breeders had a part in the improvement of the Lincoln, but probably the most constructive were the members of the Dudding family of Great Grimsby in Lincolnshire, who had been breeding Lincolns for a period of about 175 years. The Lincoln is the largest British sheep, developed specifically to produce the heaviest, longest and most lustrous fleece of any breed in the world. Great numbers were exported to many countries to improve the size and wool-quality of their native breeds. The versatile fleece is in great demand for spinning, weaving and many other such crafts.

Today, this is one of Britain's rarer breeds, categorized as being 'at risk' by the Rare Breeds Survival Trust, since there are fewer than 1,500 registered breeding females in the United Kingdom.

Mature rams weigh in at 250–350lbs (115–160kg), with mature ewes coming in at 200–250lbs (90–115kg). The fleece of the Lincoln is carried in heavy locks that often twist into spirals near the end. It has one of the longest staple lengths of all the breeds, ranging from 8–15in (203–381mm) with a yield of 65–80 per cent. Lincolns produce the heaviest and coarsest fleeces of all the longwool sheep with ewe fleeces weighing from 12–20lbs (5.4–9kg). The fleece has a numeric count of 36s–46s and ranges from 41–33.5 microns in diameter. Although coarse and somewhat hair-like, the fleece has considerable lustre.

MERINO

The Phoenicians introduced sheep from Asia Minor into North Africa, and the foundation flocks may have been introduced into Spain as late as the 12th century by the Beni-Merines, a tribe of Arabian Moors. In the 13th and 14th centuries, genetic material from England was introduced, this influence having

OPPOSITE: Like their Leicester Longwool ancestors, Lincolns grow fleeces that almost touch the ground.

RIGHT: Small quantities of Merinos were imported from Spain to the USA from 1793.

been well-documented by Spanish writers at the time. Spain became noted for its fine wool (with spinning counts between 60s and 64s) and built up a fine wool monopoly into the 16th century; wool commerce with Flanders and England was a source of income for Castile in the late Middle Ages. Most of the flocks, known as *cabañas*, were owned by the nobility or the church, the sheep grazing the Spanish southern plains in winter and the northern highlands in summer. The Mesta was an organization of privileged sheep-owners which developed the breed and controlled its migration.

An economically influential breed, the Merino has been developed because of the desirability of its fleece, and it produces some of the finest, softest wool of any sheep. Poll Merinos have no horns, and horned Merino rams have long, spiral horns that grow close to the head.

The Merino is an excellent forager and very adaptable. It is bred predominantly for its wool, and its carcase size is generally smaller than that of sheep bred for meat. The South African Meat Merino, American Rambouillet and Merinofleischschaf have been bred to balance both wool-production and carcase-quality.

The term Merino is widely used in the textile industries and is attributed various meanings. Originally it denoted the wool of Merino sheep reared in Spain, but due to the equivalent quality of Australian and New Zealand wools the term has acquired a broader use. In the dress-goods and knitting trades the term 'Merino' means an article containing Merino wool.

OPPOSITE & BELOW: Merinos, bred predominantly for their wool, produce the best fleeces of any sheep in the world.

MONTADALE

A breed developed in the 1930s by E.H. Mattingly, a Midwestern commercial lamb-buyer who had dreams of developing the ideal sheep. He had been told that the best way to start was to bring together the best characteristics of Midwestern mutton-type sheep and the big Western range sheep. The Montadale is considered a dual-purpose breed, raised for both wool and meat.

Mattingly selected the Cheviot and Columbia breeds as the basis for his project. The Cheviot is a small, hardy sheep developed in Scotland. It is known for its style, correctness and good muscling. The Columbia was developed in Wyoming and Idaho in the early 1900s, this being a large, big-bodied sheep with a heavy, good-quality fleece. The first cross of Columbia rams on Cheviot ewes proved to Mattingly that his project was on track. But he also tried breeding Cheviot rams to Columbia ewes with even more successful results and the cross was chosen as the foundation for the new breed.

The first Montadales were selected and linebred for nine years to develop uniformity in breed characteristics and type. Then, the sheep were exhibited in competitions across the country, quickly attracting the attention of progressive sheep-producers. The growth and progress of the breed has been remarkably rapid ever since.

NAVAJO-CHURRO

A breed descended from the Churra, an ancient Iberian breed. The Churra, renamed Churro by American frontiersmen, was first imported to North America in the 16th century and was used to feed Spanish armies and settlers. By the 17th century Churros were popular with the Spanish settlers in the upper Rio Grande Valley. Flocks of Churros were also acquired by Native Americans through raids and trading, and soon became an

important part of the Navajo economy and culture. Within a century, herding and weaving had become a major economic asset for the Navajo. The Churro fleece was admired for its lustre, silky feel, variety of natural colours and durability, and it was from this wool that the early Rio Grande, Pueblo and Navajo textiles were woven. Then, a series of US government-sponsored flock reductions and crossbreedings decimated the Navajo flocks until the Churro all but disappeared. Restoration of the breed began in the 1970s when breeders began acquiring Churro phenotypes with the purpose of preserving the breed and revitalizing Navajo and Hispanic flocks. The Navajo-Churro Sheep Association was formed in 1986 to preserve and promote the breed, the name 'Navajo-Churro' having been chosen to indicate both Navajo and Spanish heritage. While the breed is no longer in danger of extinction, Navajo-Churro sheep are still considered to be rare.

OPPOSITE & ABOVE: The Navajo-Churro stems from an ancient Iberian breed that was brought from Spain to North America.

The breed is renowned for its hardiness and adaptability to extremes of climate. It is raised primarily for its wool, the fleece consisting of a protective topcoat and soft undercoat. Some rams have four fully-developed horns, a trait shared with few other breeds in the world. The Navajo-Churro has also gained popularity due to its reputation for low-maintenance, resistance to disease, and lean meat; some even say it is a handsome animal. Ewes often bear twins.

Colour is separated into fleece colour and points colour (leg/head), and includes reds, browns, black, white and mixes. There may also be different colour patterns.

NEWFOUNDLAND

Also called the Newfoundland Local, this is a breed of sheep native to Newfoundland. The island of Newfoundland is mainly supported through fishing, but a persistent population of sheep has been present since the 16th and 17th centuries, and a native landrace has developed since that time. Although it has never been fully recognized as a breed, the Newfoundland sheep has a distinct genetic base. Both Border Cheviot and North Country Cheviot are known to have been major contributors to Newfoundland bloodlines, and there may also have been contributions from other breeds, such as Suffolk, Border Leicester, Dorset and Scottish Blackface.

Newfoundland sheep show fairly wide variability but are generally small with white or off-white coats. Very few ewes are horned, and rams

can be either horned or polled. All Newfoundlands are characterized by their hardiness and ability to survive on poor native forage.

Coinciding with a general decline in sheep populations on the island, Newfoundland sheep of the native type declined to fewer than 200 individuals at one point. Thanks to attention from universities, rare-breed organizations and farmers, numbers have increased, but the breed is still considered to be critically endangered.

NORTH COUNTRY CHEVIOT

A type of Cheviot, North Country Cheviots are bred predominantly in Scotland but also appear in other parts of the United Kingdom, including Northern Ireland.

In 1791, Sir John Sinclair brought ewes from the Cheviot Hills, near to the English border with Scotland, to the counties of Caithness and Sutherland in northern

OPPOSITE: The Newfoundland sheep is native to the island of Newfoundland.

RIGHT: North Country Cheviots are predominantly Scottish sheep.

Scotland. He named these sheep 'Cheviots', after the hill area where they originated. Another hill breed was introduced into the ranges of central Scotland, thus the Scottish Blackface created a definite separation between the northern counties of Caithness and Sutherland and the border region in southern Scotland. Most authorities speculate that both English and Border Leicesters may have been introduced into the North Country Cheviots at this time. The result was a larger sheep with a longer fleece that matured earlier. The North Country is about twice the size of its southern relative.

In 1912, Caithness and Sutherland breeders formed the North Country Sheep Breeders Association to manage shows and sales. In 1945, the organization was reformed into the existing North Country Cheviot Sheep Society for registration, exporting, promotion and improvement of the breed.

OXFORD

Also known as the Oxford Down, this is an English breed developed in the 1830s by crossing Cotswolds with forerunners of the Hampshire and

using the resulting crossbreds to form the basis of the present-day breed. It is a breed primarily raised for meat.

The Oxford is short, relatively large-bodied and hornless, with a brown face and legs covered in light-brown wool. The Oxford produces the heaviest fleece of any of the Down breeds. The breed's capacity to produce a large, meaty carcase for further processing has stimulated interest from the meat industry, and it also grows the most wool of any of the terminal sire breeds.

Mature weights for rams range from 200–300lbs (90–136kg), with ewes weighing in at 150–200lbs (68–90kg). Fleeces from mature ewes weigh between 8–12lbs (3.6–5.4kg), with a fibre diameter of 30–34.5 microns and a numerical count of 46–50. The staple length of the fleece ranges from 3–5in (76–127mm) and has a yield of 50–62 per cent.

PANAMA

A breed native to the United States. Although the name suggests it is a native of Panama, the breed was in fact named for the Panama-Pacific International Exposition, where it was shown early on in its history.

The Panama is rare, largely unknown outside of Idaho and Montana, and is one of only a few breeds that were developed by private individuals in the United States. The Panama was developed by James Laidlaw, a Scottish immigrant to the region who was seeking a large sheep better suited to the range than the Merinos, which were most common in the late 19th and early 20th centuries. The foundation of the breed was Rambouillet rams crossed to Lincoln ewes.

Panamas are polled sheep with white fleeces of medium length, and they have large bodies suitable for

OPPOSITE & ABOVE: Oxfords are well-muscled sheep with good-quality fleeces.

meat production. They are particularly known for their tolerance of the rough conditions of the northern Rocky Mountains. A breed registry was formed in 1951, but it has been inactive and it is unclear whether or not the majority of Panamas present today have been crossbred. The sole remaining flock that is known to be pure is the one maintained by Idaho University.

PERENDALE

Developed in New Zealand by Massey Agricultural College (now Massey University) for use in steep hill situations, the Perendale is named after Sir Geoffrey Peren, and it achieved its aim, as the offspring of Romney ewes and Cheviot rams, by having sturdy legs.

Since the early 1980s the flock numbers of this sheep have become reduced, mainly because hill-country farming has diminished, there has been a lower demand for medium-coarse wool, and because modern Romneys are more adaptable to the terrain.

The Perendale is a dual-purpose sheep producing wool with a fibre diameter of 29–35 microns and with a 5-in (127-mm) staple length. Wool grade is 44s–54s.

The Perendale is a characteristically high-fertility animal, with great potential to produce a prime ewe lamb when crossed with the Merino. As a purebred, its hardiness makes it ideally suited to colder, high rainfall areas. The Perendale is easy to care for; the ewes have little trouble lambing and are good mothers. Mature rams weighs in at around 220–260lbs (100–118kg), the ewes coming in at 120–150lbs (54–68kg).

POLYPAY

A white, medium-sized sheep which was developed in the 1960s at the US Sheep Experiment Station in Dubois, Idaho. In general, Polypay sheep are noted for being a highly prolific, dual-purpose breed. Dr. C.V. Hulet set five main goals for the new breed: high lifetime prolificacy, a large lamb crop at one year of age, the ability to lamb more frequently than once a year, rapid growth-rate of lambs, and desirable carcase quality.

The original breeding stock consisted of Finnsheep, for their high prolificacy, early puberty and short gestation; Dorset, for their superior mothering ability, carcase quality, early puberty and long breeding season; Targhee, for their large body size, long breeding season and quality fleeces; and Rambouillet for their adaptability, hardiness, productivity and quality fleeces.

In 1968 the first crossbreeding was performed, the result having been produced by 1970. The Polypay name was created in 1975 from ‘poly’, meaning multiple, and ‘pay’, meaning return on investment. The American Polypay Association was formed in 1980.

RAMBOUILLET

Also known as the Rambouillet Merino or the French Merino, the development of the Rambouillet breed started in 1786 when Louis XVI purchased over 300 Spanish Merinos (318 ewes, 41 rams, 7 wethers) from his cousin, Charles III of Spain. The flock was subsequently developed on an experimental royal farm, the Bergerie Royale (now Bergerie Nationale) owned by Louis XVI and built on his domain of Rambouillet,

30 miles (50km) south-west of Paris. The flock was raised exclusively at the bergerie, with no sheep being sold out for many years.

Outcrossings with English longwool breeds, as well as selection, produced a well-defined breed differing in several important points from the original Spanish Merino: the size was greater, with full-grown ewes weighing up to 200lbs (90kg) and rams up to 300lbs (136kg) live weight. The wool clips were larger and the wool-length had increased to more than 3in (76mm).

In 1889, the Rambouillet Association was formed in the United States with the aim of preserving the breed. It has been estimated that 50 per cent of the sheep on the US Western ranges are of Rambouillet bloodstock. Rambouillets have also had an enormous influence on the development of the Australian Merino industry through Emperor and Peppin Merino studs.

The breed is well-known for its wool but also for its meat in the form of both lamb and mutton. It has been described as a dual-purpose breed, with superior wool and near-mutton breed characteristics. The fleece was valuable in the manufacture of cloth, at times being woven into a mixed fabric of cotton warp and wool weft, known as *delaine*.

RIDEAU ARCOTT

A breed of domestic sheep native to Canada, the Rideau Arcott is one of only a few livestock breeds native to that country, the name 'Rideau' being common in Ottawa. The latter half of its name is an acronym for the Animal Research Centre in Ottawa, where it was developed along with the Canadian Arcott.

The Rideau Arcott is a synthesis of many different breeds, but is primarily an amalgam of Finnsheep (Finnish Landrace), Suffolk, East Friesian, Shropshire and Dorset Horn. The research flock was closed in 1974, and the breed was distributed to shepherds beginning in 1988.

Today, the Rideau Arcott is a maternal breed, with the birth of twins and triplets being the norm. Crossbreeding with Rideau rams is said to increase lambing rates to 180 per cent or more.

ROMANOV

A breed originating in the Volga valley north-east of Moscow, Russia. These domestic sheep got the name 'Romanov' from the old royal family of Russia. Soon after they were noticed in the 18th century, the sheep were imported into Germany and then into France. In 1980, 14 ewes and four rams were bought by the Canadian government and were quarantined for 5 years. After testing, some of the Romanov breed was brought into the United States. The distribution of this unique breed is worldwide, and it is raised primarily for meat.

Romanovs are one of the Northern European short-tail breeds. They are pure black when they are born, but as they grow older the colour quickly changes to grey. The average weight of a mature male is 120–176lbs (54–80kg) with females coming in at 88–110lbs (40–50kg).

Romanovs are sexually mature by 3–4 months and will begin breeding in any month of the year. Ewes are capable of producing quadruplets, quintuplets, and sometimes even sextuplets all at one time. British and North

American breeds of domesticated sheep are genetically different because the Romanov is a 'pure-gene' breed rather than a 'cross', even though they are often crossed with more popular breeds to increase their prolificacy.

ROMELDALE/CVM

The Romeldale is an American fine-wool breed, and the California Variegated Mutant, or CVM, is its multi-coloured derivative. Both the Romeldale and CVM are unique to the United States and are on the American Livestock Breeds Conservancy's critical list.

The Romeldale began in California when in the early 1900s A.T. Spencer purchased the entire contingent of New Zealand Romney rams that were exhibited at the 1915 Pan American Exposition in San Francisco. He bred these rams to his Rambouillet ewes, the intention being to improve the meat and wool qualities of his stock. These Romney-Rambouillet crosses were developed for several years and eventually becoming known as Romeldales. Much of the establishment of the Romeldale breed was accomplished by the Sexton family during the 1940s and 1950s, with sheep selected for high rates of twinning, maternal ability, and non-seasonal reproduction. Soft-handling wool was also a priority, as was fleece weight (10–15lbs/4.5–7kg) with a spinning grade of 60s to 64s.

The California Variegated Mutant or CVM is a rare sub-type of the Romeldale breed known for its unusual colour. The majority of sources (as well as the breed association) refers to the two together as Romeldale/CVM, and they are generally considered to be two types of the same breed. Other than colour, the two share most of the same physical and temperamental traits, being medium to large sheep at 150–275lbs (68–125kg) in weight. The original Romeldales are mostly white, through any sheep not conforming to the CVM standard may be registered as Romeldale. CVMs are all natural-coloured sheep with a badger face pattern. Unusually for such sheep, their fleeces tend to darken with age rather than grow lighter. Both Romeldales and CVMs are polled. The soft wool and unusual colours of the CVM are especially prized by hand spinners.

ROMNEY

The Romney and Leicester breeds are early examples of sheep evolving from medieval longwool types. The sheep recognized by 1800 as 'Romney Marsh' or 'Kent' were improved in body type and fleece quality through crossings with Bakewell's English Leicester. Exported to other continents, the Romney is an economically important sheep, especially in the meat and wool export trades of New Zealand.

OPPOSITE: The CVM sheep is recognized as a colour variety of the Romeldale composite breed and is not a unique breed in its own right.

ABOVE RIGHT: The Romney fleece hangs in separate, lustrous locks. It is also high-yielding and easy to spin.

In 1904, William Riddell & Sons of Monmouth, Oregon, imported the first Romneys to North America, where their popularity increased rapidly. The American Romney Breeders Association was founded in 1912 by Joe Wing, a world traveller and a great judge of sheep. At one

time with the University of California, he was an early secretary of the ARBA, and his expertise was instrumental in the development of Romneys in America.

The breed developed in Kent and East Sussex in and around the area of low-lying grassland, known since the Middle Ages as the Romney Marsh, this being the place where, for many centuries, sheep had been brought to graze from other parts of England to be fattened for market. Although windy in winter and fever-ridden, the marsh consists of rich pastures superimposed upon estuarine sediments. It was originally a tidal salt marsh, but it was 'inned' (protected from the sea by dykes) between 1150 and 1400 to form non-tidal land.

The Romney is in general an open-faced breed with long wool that grows down and over the legs. Romney breed standards are not identical across all countries but they have much in common. The oldest

Romney breed society, being that of England, founded in 1895, in 1991 set the standard as follows. Head wide, level between ears, with no horns nor dark hair on the poll. Eyes should be large, bright and prominent and the mouth sound. Face in ewes full, and in rams broad and masculine in appearance. Nose and hooves should be black. Neck well set in at the shoulders, strong and not too long. Shoulders well put in and level with the back. Chest wide and deep. Back straight and long, with a wide and deep loin. Rump wide, long and well-turned. Tail set almost even with the chine. Thighs well let down and developed. The face should be white, and the skin of a clean pink colour. Ribs should be well-sprung. Legs well-set, with good bone and sound feet. Sheep should stand well on their pasterns. The fleece should be of a white colour, with even texture and a good decided staple from top of head to end of tail and free from kemp.

It is the low grease content of Romney wool that makes it a very light-shrinking fleece upon washing, and it is consequently high-yielding with a range of between 65–80 per cent. The long, lustrous fleece, hanging in separate locks, also makes it especially attractive to hand spinners, being easily spun in the grease or after washing and carding, and it readily takes dye. The natural black, grey, silver and brown colours frequently found in Romneys are among the most sought after at shows and sales, and a fleece of hand-spinning quality may often sell for three to ten times the amount a commercial buyer would pay for it.

OPPOSITE & ABOVE RIGHT: Romney sheep are named for England's Romney Marsh, where they have grazed since medieval times.

ROYAL WHITE

A new breed of domestic sheep in the United States, privately funded and developed by William Hoag for high performance and low maintenance. This is a hybrid breed in which the best traits of St. Croix and Dorper have been used and their negative traits removed via selection over many years. It has resulted in a sheep that produces more meat, less fat and less wool to optimize the meat per feed/pound ratio. Lean, tender meat, with a mild, sweet flavour is the purpose of this breed, and it is also highly disease-resistant. Royal Whites outgain goats and will convert feed better to gain than, for example, Angus cattle. They also consistently outperform both Dorper and Katahdin sheep (wool-hair types).

Royal Whites grow a longer hair coat in the fall, shedding it off naturally in the spring. The coat is pure white. Both rams and ewes are naturally hornless.

SANTA CRUZ

An extremely rare breed that once existed as a feral population on Santa Cruz Island of the Californian Channel Islands. The sheep is primarily a wool breed.

In the mid-1800s, sheep that were most likely of the Merino, Rambouillet (French Merino) or Churra breed were brought to Santa Cruz, and by the 1860s thousands of sheep were grazing freely on the island. Throughout the 20th century, ranching declined and most of the sheep became feral. In 1978, when sheep numbers were estimated to be over 20,000, Nature Conservancy gained control of the island and, together with the National Park Service, began to kill or remove all sheep remaining in order to prevent overgrazing of the island's vegetation. Today, the breed has fewer than 200 animals remaining, and is it listed as 'critical' by the American Livestock Breeds Conservancy. A small population of the sheep exists on the mainland, having been largely placed there through adoption.

BELOW & OPPOSITE: Scottish Blackfaces are slow-maturing animals, but they have long productive lives. Their wool is popular with hand spinners.

Santa Cruz sheep are relatively small and extremely hardy: they are good foragers and need no assistance with lambing. Because of the finewool breeds from which they are derived, Santa Cruz sheep are unique among formerly feral island breeds in that their medium to fine wool has a soft feel. Fleeces are mostly white, but some coloured sheep appear from time to time.

SCOTTISH BLACKFACE
Also known as the Blackfaced Highland, Kerry, Linton, Scottish Mountain, Scottish Highland, Scotch Blackface and Scotch Horn,
the origins of the breed are uncertain, but it was developed on the Anglo-Scottish border, it being unclear when it became a distinct breed. Early records show that monks in the 12th century raised sheep that may be the progenitors of the modern Scottish Blackface breed. The monks used the wool of the dun-faced sheep, as they were often called, for their own clothing and exported large amounts of it to Europe. Later records show that in 1503 James IV of Scotland established a flock of 5,000 Scottish Blackface sheep in Ettrick Forest, in the area south of Peebles in the Border country.

Today, this is the most numerous breed in the British Isles, with roughly 30 per cent of all sheep in the UK being Scottish Blackface. It epitomizes the mountain sheep in that its long, coarse wool shields it from the rain and biting winds, enabling it to survive the harshest winters in the most extreme parts of Britain, from the Scottish Highlands to the wilds of Dartmoor. Blackfaces are horned in both sexes and, as their name suggests, usually have black faces (but sometimes with white markings), and also black legs. It is a breed primarily raised for meat.

Several types have evolved over the years, the most common being the Perth variety, which is large-framed with a longer coat, and mainly found in north-east Scotland, Devon, Cornwall and Northern Ireland; the medium-framed Lanark type, with shorter wool, is commonly found in Scotland and Ireland.

The introduction of black-faced Highland sheep to America first occurred in June 1861, when Hugh Brodie imported a ram and two ewes for Brodie & Campbell's New York mills. In 1867 the flock and increase were purchased by T.L. Harison of Morley, St. Lawrence County, while Isaac Stickney, of New York, also imported a small flock in that year for his farm in Illinois.

Artisans have long used the horns of Blackfaces for the carvings on shepherd's crooks and walking sticks. In the US the fleeces are becoming of interest to hand spinners for use in tapestries and the making of rugs and saddle blankets.

BELOW & OPPOSITE: Shetland sheep are small, thrifty, slow-growing and long-lived and have retained many of their primitive survival instincts.

SHETLAND

A small, finewool breed originating in the Shetland Islands of northern Britain, but now kept in many other parts of the world. Shetlands are one of the Northern European short-tails, a group which also includes the Hebridean, Soay, Finnsheep, Norwegian Spelsau, Icelandic, Romanov and others. The Shetland is classed as a landrace or 'unimproved' breed. It is raised primarily for meat.

Although Shetlands are small and slow-growing, compared with commercial breeds, they are hardy, thrifty, easy lambers, adaptable and long-lived. The Shetland breed has survived for centuries in difficult conditions and on a poor diet, and consequently thrive when exposed to better conditions. However, they retain many of their primitive survival instincts, making them easier to care for than many modern breeds.

The wool produced by Shetland sheep is also a valued commodity. Shetlands appear in a wide variety of colours, many of which have been given traditional names by breeders, and they are commercially important to a wool industry where natural wools are often used undyed. Tweed is also produced from the coarser Shetland wool but the islands are best-known for their multi-coloured knitwear and for the traditional knitted lace shawls; these are so fine that they can be passed through a wedding ring. Fleeces usually weigh between 2 and 4lbs (0.91 and 1.8kg).

SHROPSHIRE

A breed originating in the hills of Shropshire and North Staffordshire, England, during the 1840s. Breeders in the area used the local horned black-faced sheep and crossed them with a few breeds of white-faced sheep (Southdown, Cotswold and Leicester). This produced a medium-sized polled (hornless) sheep that produced good wool and meat. In 1855 the first Shropshires were imported into the United States (Virginia). This breed is raised primarily for meat.

The Shropshire was officially recognized as a distinct breed by the Royal Agricultural Society in 1859. The popularity of the Shropshire breed grew rapidly in England, and in 1882 Shropshire breeders founded the Shropshire Sheep Breeders' Association and Flock Book Society, the world's first such society for sheep. The same year the society published the first flock book, this being a record of sheep bred and their breeders. The society still survives, publishing a flock book annually.

By 1884 more Shropshires were being exhibited at local shows than all other breeds combined. The first documented flock in the United States (one ram and 20 ewes) was brought to Maryland in 1860 by Samuel Sutton. In 1884 the American Shropshire Registry was formed and by the turn of the century the Shropshire was the most numerous sheep breed in the United States. Thousands of Shropshires were exported to the United States after that, as well as to other parts of the English-speaking world, notably Australia and New Zealand, and to

Shropshires are gentle in disposition, making them perfect for the small farm or as a club project for children.

South America. The breed's adaptability to most environments, and their dual-purpose function, led quickly to them becoming a popular breed.

By the 1930s, in the USA, the Shropshire had been dubbed 'the farm flock favourite'. But in the 1940s US breeders began producing Shropshires with more wool-cover and decreased size. This led to the breed having increased wool around the eyes, and thus necessitated trimming away for better sight. This drawback, and the overall loss of size, led to the numbers of the breed decreasing among American farmers. They were no longer the most popular breed of sheep, and consequently became increasingly rare around the world and even in their native land.

SOAY

A primitive breed (*Ovis aries*) descended from a population of feral sheep on the 250-acre (1-km^2) island of Soay in the St. Kilda archipelago, about 40 miles (65km) from the Western Isles of Scotland. Undiluted by interbreeding, they are a genetic archive of the Neolithic origins of domesticated sheep, and they are one of the Northern European short-tail breeds. The name 'St Kilda' does not refer to the Soay sheep, but is a former name for another short-tailed type, the Hebridean, that was raised primarily for meat.

Soays are similar to other wild ancestors of domestic sheep, such as the Mediterranean mouflon and the horned urial sheep of Central Asia. Their more immediate origins are uncertain, however, it being unclear whether they came to the island some time during the Bronze Age, or were brought by Vikings in the ninth and tenth centuries, the name of the island, Soay, being Old Norse for 'Island of Sheep'.

They are much smaller than modern domesticated sheep but hardier. Soays are extraordinarily

ABOVE LEFT: Shorn Shropshires.

OPPOSITE, PAGES 236 & 237: The Soay is truly unique, being a primitive breed with Neolithic origins.

agile, and tend to take refuge among cliffs when frightened. Coat colours tend to be either blonde or dark brown with a buffish-white underbelly and rump (known as *lachdann* in Scottish Gaelic, which equates with the Manx *loaghtan*), or totally black or fawn-coloured, while a few have white markings. In the early 20th century, some Soay sheep were translocated to establish exotic flocks, such as that of Park Soay at Woburn Abbey, established by the Duke of Bedford in 1910, and selected for their 'primitive' characteristics. A number of Soay sheep were also taken from Soay to the island of Hirta by the Marquess of Bute in the 1930s, after the human population was evacuated.

The Hirta population has been the subject of scientific study since the 1950s, and makes an ideal model subject for scientists researching evolution, population dynamics and demography. This is because the population is unmanaged, closed

(no emigration or immigration) and has no significant competitors or predators.

This breed has an extremely fine fleece and, in contrast with the mouflon, the inner fleece is highly developed, making it difficult to distinguish an outer coat. This is a clear indication that the Soay is indeed the product of a domesticated breed in prehistoric times. The breed also lacks the flocking instinct of many other sheep, and attempts to work them using sheep dogs result in a scattering of the group.

SOUTHDOWN

A small, dual-purpose British sheep raised principally for meat. The Southdown breed was developed by John Ellman of Glynde, near Lewes, East Sussex about 200 years ago. His work was continued by Jonas Webb of Babraham in Cambridgeshire, who developed the larger animal that we see today. It was exported to New Zealand and was used in the breeding of Canterbury Lamb. The Southdown was involved in crossbreeding to develop other breeds, i.e., with existing stock, the Hampshire; via the Hampshire, the Oxford Down; with the Norfolk Horn, the Suffolk.

In Britain, the Southdown is recognized by the Rare Breeds Survival Trust as a native breed, although today it is more popular with the smaller-scale breeders of sheep.

It has been split into two sub-breeds. Today, the Southdown raised by commercial growers is larger than the 'traditional' Southdown of the past. North American Southdowns are also taller than their British counterparts. The original bloodlines of the English Southdown are in the 'Baby Doll' Southdowns in the US, selected specifically for their smaller size, the focus being on wool and hobby breeding rather than on commercial meat production. Baby Doll breeders claim that their sheep are closer to the old, traditional British

Southdown than are the commercial Southdowns grown today.

Mature rams weigh in at 190–230lbs (86–104kg), with the ewes coming in at 130–180lbs (59–82kg). Fleece weights for mature ewes are between 5 and 8lbs (2.25 and 3.6kg) with a yield of 40–55 per cent. The fleeces are considered to be medium wool type with a fibre diameter of 23.5–29 microns and a numerical count of 54–60. The staple length ranges from 1.5–2.5in (38–63.5mm).

ST. CROIX

A breed native to the US Virgin Islands and named for the island of St. Croix, the sheep is also referred to as the Virgin Island White, because those that were imported into North America were selected for their white coloration.

A breed of hair sheep which does not grow wool, the St. Croix is believed to be descended from African sheep that were brought to the Caribbean on slave ships. It is a hardy tropical breed known for its resistance to parasites, and is raised primarily for meat production. Breeders have crossbred the St. Croix with others to introduce these important traits into their bloodlines. It is the foundation breed for the Katahdin and Royal White breeds.

Most St. Croix sheep are white with others being solid tan, brown, black or white with brown or black spots. Both ewes and rams are polled, and rams have a large throat ruff. Mature ewes weigh in at 150lbs (68kg), with rams coming in at 200lbs (90kg). Birth weights average 6–7lbs (2.7–3.1kg). The ewes can breed back a month after lambing and can produce two lamb crops per

year. Twins are common, while triplets, and occasionally quadruplets, are often produced.

SUFFOLK

Evolved from the mating of Norfolk Horn ewes with Southdown rams in the Bury St. Edmunds area of Suffolk, England, Suffolk sheep were known as Southdown Norfolks or locally as 'Blackfaces'.

First mention was made of the breed in 1797 when, in his 'General view of agriculture in the county of Suffolk', Arthur Young remarked: 'These ought to be called the Suffolk breed, the mutton has superior texture, flavour, quantity and colour of gravy.' The breed was first exhibited at the Suffolk Show in 1859, and the first flock book was published in 1887. This contained 46 flocks ranging in size from 50 to 1,100 ewes and averaging 314. All 46 flocks were in East Anglia and 34 were in Suffolk itself. The oldest was that of E.P. & H. Frost of West Wratting, established in 1810.

LEFT & OPPOSITE: The St. Croix was the foundation of the Katahdin and Royal White breeds.

Suffolks developed around the rotational system of farming in East Anglia, grazing on grass or clover in the summer. After weaning, the ewes would be put on salt marshes or stubbles. Swedes, turnips or mangels were grazed in the winter in a very labour-intensive system, with a fresh area fenced off each day. Lambing was outdoors in the fields in February or March, with a hurdle shelter or in open yards surrounded by hurdles and straw.

The breed expanded rapidly, with the first flock in Ireland established in 1891, in Scotland in 1895 and in 1901 in Wales. From the earliest days sheep were exported around the world to Austria, France, Germany, Switzerland, Russia, North and South America and the colonies.

Originally renowned as a producer of mutton, the Suffolk has developed over the years to match consumer demands. Suffolks are now found throughout the world's sheep-producing countries. They are the flag-ship breed in the British Isles and are recognixed as the leading terminal sire on a variety of ewes to produce top-quality prime lamb.

TARGHEE

A breed developed in the early 20th century by the USDA's Agricultural Research Service. Targhee sheep are a dual-purpose breed, with heavy, medium-quality wool and good meat production characteristics. They are hardy and especially suited to the ranges of the West where they were developed. Targhees are especially popular in Montana, Wyoming and South Dakota, where they are raised primarily for wool.

The breed was named after the Targhee National Forest which surrounds the US Sheep Experiment Station in Idaho. The breed's ancestors were Rambouillet, Corriedale and Lincoln sheep. Though development of this breed for the Western ranges of the US began as early as 1900, the flock book was closed in 1966, which means that only the offspring of registered Targhees can themselves be registered.

TEESWATER

Coming from Teesdale in Yorkshire, in the north of England, the Teeswater is a longwool sheep. It was bred by farmers for about 200 years, but by the 1920s had become rare. The breed has fortunately enjoyed a renaissance since the Second World War, and can be found all over the UK, although the Rare Breeds Survival Trust categorizes the breed as vulnerable. The Teeswater is raised primarily for meat.

The Teeswater Sheep Breeders' Association was formed in 1949 with the aim of encouraging and improving the breeding of Teeswater sheep and to maintain their purity. It particularly wished to establish the supremacy of Teeswater rams for crossing with hill sheep of other breeds for the production of halfbred lambs.

The wool of the Teeswater is fine, long-stapled and with high lustre, with each lock hanging free and with no tendency to felt. There must be no dark fibres in the fleece, which should be uniform in texture over the whole body. The Teeswater produces a kemp-free fleece, a characteristic that it passes on to its offspring.

TEXEL

A breed originating in Texel, the largest of the Friesian Islands off the north coast of the Netherlands. The exact origin of the breed is unknown although it is thought to be a cross of multiple English breeds.

It was slowly evolved into a meat breed of outstanding carcase quality, and is now one of the most common meat breeds in the Netherlands, where it makes up 70 per cent of the national flock. In 1985, the first Texels in the United States were imported by the Meat Animal Research Center at Clay Center, Nebraska. In 1990, and after a five-year quarantine, some were released for purchase by private individuals. Private importations have since been made by a handful of breeders in the United States. It is also popular in Australia, New Zealand, Uruguay and the rest of Europe.

The Texel is a heavily-muscled sheep, passing this quality on to its crossbred progeny. The wool is around 32 microns and is mostly used for hosiery yarns and knitting wools.

TUNIS

A breed with a narrow head with pendulous ears. Tunis lambs are robust at birth and are protected by a double coat, which is mahogany red on the surface. White spots on the top of the head and tip of the tail are common. The creamy-white fleece appears as the lamb matures. The wool is lustrous and long-stapled, being 4–6in (102–152mm) in length. Tunis ewes are heavy milkers, and twin lambs are commonly born. Ewes are known to breed out of season, which makes them valuable for autumn lamb production. Tunis sheep are also known for disease-resistance and the ability to tolerate both warm and cold climates. Their meat is tender and flavourful but without the strong mutton taste.

The Tunis is one of the oldest breeds, being descended from ancient fat-tailed sheep referred to in the Bible. As the name indicates, the Tunis originated in Tunisia on the

The Texel is a relative newcomer to North America, being the only Dutch breed presently on the continent.

coast of North Africa. The earliest documented importation to the USA occurred in 1799, being a gift from the ruler of Tunisia and entrusted to the care of Judge Richard Peters of Pennsylvania. One of the largest advocates of the Tunis breed was Thomas Jefferson, who owned a fairly large flock of sheep. The popularity of the breed spread quickly in the US and flocks were established on the east coast and New England, where many remain today. Because of their delicious meat, most of the southern flocks were wiped out during the Civil War. The National Tunis Sheep Registry, Inc. has experienced continuous growth in registrations and transfers, moving the breed from its threatened status to the watch list of the ALBC Conservation Priority List.

WENSLEYDALE

A breed that originated in the Wensleydale region of North Yorkshire, England. The breed was developed in the 19th century by crossing English Leicester and Teeswater sheep. One of the largest and heaviest of all sheep breeds, the Wensleydale has long, ringlet-like locks of wool. It is categorized as being 'at risk' by the Rare Breeds Survival Trust of the UK as it has fewer than 1,500 registered breeding females. Rams are predominantly used to cross with other breeds to obtain market lambs and high-quality wool.

The Wensleydale is a large longwool sheep with a distinctive deep-blue head, ears and legs. Wool from the breed is acknowledged to be the finest lustre longwool in the world. Fleeces from purebred sheep are considered to be kemp-free.

WILTSHIRE HORN

Originating in Wiltshire in southern England, the Wiltshire Horn is raised for meat. The breed has an unusual feature in that it sheds its hair in spring, making shearing unnecessary. Ewes are good mothers, have high fertility and, along with the rams, are relatively intelligent compared with most other sheep. The fact that they do not require shearing or crutching, and are not readily affected by fly strike, makes them useful for lamb production.

Although the breed was abundant in the 1700s and 1800s, it was nearly extinct at the start of the 1900s. In 1923, in an attempt to save the breed, the Wiltshire Horn Breed Society was formed. By the early 1980s, there were 45 registered flocks in the UK.

The Wiltshire Horn is one of the foundation breeds for Katahdin and Wiltipoll sheep. Both males and females are horned and rams' horns grow one full spiral each year until maturity. Both sexes are white with occasional black spots on the undercoat. This is a hair breed, growing a thick, coarse coat in the winter and shedding it in spring. Rams weigh about 250lbs (114kg) and ewes 150lbs (68kg).

OPPOSITE: Wensleydales have finer fleeces than most longwool breeds.

BELOW: Wiltshire Horns shed their hair in spring, making shearing unnecessary.

CHAPTER FIVE
HORSES

BRABANT

The Brabant, or Belgian Heavy Draft Horse, comes from the area of Belgium that has Brussels as its capital. It is of ancient origin, only slightly more recent than the Ardennais to which it owes part of its lineage: the other part of its inheritance is thought to have stemmed from the Flanders Horse of the 11th–16th centuries, which in turn is believed to be descended from the ancient horses of the Quaternary period. For centuries, Belgian breeders produced their stock by selective breeding, which also included inbreeding.

The Brabant's very existence is a direct result of the geology of the area; the rich heavy soil required a horse with great pulling power and big strong joints to enable it to lift its huge feet out of the thick clods of mud. As a result, three distinct bloodlines emerged 100 years ago which intermingled to create the modern Brabant: the Gros de la Dendre, which was muscular and strong with huge legs; the Gris de Nivelles, with good conformation and a certain elegance; and the Colosse de la Mehaigne, which was large and had a lively temperament.

Over the centuries, the Brabant has had enormous influence on today's modern breeds, much in the same way as the Arab bloodline has been added to improve existing stock. In the Middle Ages the horse was imported all over Europe and its bloodlines are present in the German warmbloods. The Russians introduced native breeds to it to produce working horses and its influence is also present in the Shire, Irish Draught and Clydesdale, to name but a few. Today, Brabants are still part of the foundation stock for the breeding of warmbloods. They now appear throughout the world where they are still used in agricultural work, logging and as dray horses. They can also be seen in the show ring.

The head is fairly square with a straight profile, small pricked ears and deep-set eyes with a kindly expression. The neck is short and very strong and set high with a large crest. The shoulders are sloping and the chest is wide and deep. The body is short with a well-muscled back and strong quarters. The legs are fairly long and muscular and the hooves are large, rounded and tough; there is not much feathering present.

The Brabant is an extremely docile animal, to the point that it could almost be described as sluggish. However, it has an equable nature, is obedient, and possesses pulling power equal only to the Shire's, for which it is highly prized. It is a hard worker with plenty of stamina and a strong constitution, requiring relatively little food for its size. Commonly a light chestnut with a flaxen mane, Brabants can also be red roan, bay, dun and grey. Height is 16.1–17hh.

The Brabant is an ancient horse from the former duchy of that name in Belgium.

CLEVELAND BAY

The excellent Cleveland Bay is Britain's oldest breed and dates back to medieval times. Gradually, however, it became rarer, and numbers dipped to a critical level in the last century. Thankfully, it is once more gaining in popularity and numbers have begun to increase.

The breed is related to the Chapman Horse, which lived in north-east Yorkshire in the Middle Ages, and which received Iberian and Barb bloodlines. Clevelands were then used mainly as packhorses and for agricultural work where they were greatly admired for their strength and ability to carry heavy loads for long distances. The name comes from the area (Cleveland) where they were bred and the fact that their colour is bay.

Later the breed was crossed with Thoroughbred to produce a lighter, elegant carriage horse which is a feature of the Cleveland Bay today. Sadly the previous type is now extinct.

In a previous age the Cleveland Bay was very popular, but the development of motorized

BELOW: The Cleveland Bay is a rare breed which is recovering its lost popularity.

OPPOSITE: Bred as a draught horse, the Clydesdale is nevertheless a very refined animal.

transport saw its demise and by the 1970s the breed had been reduced to an all-time low. In some respects, however, they are still very much in evidence; Cleveland Bays have been kept at the royal mews since King George V introduced them, and the Hampton Court Stud still actively breeds them for state and ceremonial occasions.

Today Cleveland Bays and part-breds can be seen in showjumping, dressage, eventing, driving and hunting, where they are admired for their surefootedness and great stamina.

Clevelands are calm and intelligent and seemingly possess the ability to think for themselves. They are honest, strong and confident with enormous powers of endurance. They are exclusively bay, with a rich black mane and tail and black-stockinged legs with no white. Height is 16–17hh.

CLYDESDALE

The establishment of the Clydesdale began in the late-17th century when Lanarkshire farmers and various dukes of Hamilton supposedly imported Flemish

stallions, ancestors of the Brabant, into Scotland. The farmers were skilful breeders and mated them with native heavy draft mares already in existence; over the next 100 years or so, English Shire, Friesian and Cleveland Bay blood was also added. The result was the Clydesdale and it became highly prized as a draught horse. The Clydesdale Horse Society was established in 1877, almost a century and a half after the breed first began to evolve.

The breed soon became popular as a general farm horse and also for haulage over long and short distances; Clydesdales could be found in most major cities of Scotland, the North of England and Northern Ireland, as well as in agricultural areas. In fact, the horse became popular the world over, when considerable numbers were imported to North America, Canada and Australia.

The Clydesdale differs from most heavy draught horses, which tend to be squat and plain-looking; in fact, with its short-coupled body, long legs, and high head-carriage it looks positively refined.

As with all heavy horses the Clydesdale breed began to decline with the development of motorized transport and reached an even lower ebb in the 1960s and 70s. However, a few families kept the breed going and today numbers have increased though the horse is still classified as 'at risk' by the UK's Rare Breeds Society. Today they are highly valued in the show ring as well as in harness and as dray horses, where they take part in displays and are even used to pull wedding carriages.

The head is proudly held with medium, well-shaped ears which are pricked and alert; the eyes are kind and intelligent. The nose is slightly Roman and the nostrils large. The neck is long and well-set, with a high crest leading to high withers. The back is slightly concave and short and the quarters are well-developed and powerful. The legs are straight and long with plenty of feathering. The feet are large and require careful shoeing as contracted heels have a tendency to develop.

These charming horses are energetic with an alert, cheerful air. They are even-tempered and enjoy the company of other horses and human beings. They are extremely strong with a lively action and a slight tendency to dish.

Clydesdales can be bay, brown and black and usually have white patches all the way up the legs and under the belly, which can turn roan in places. Height is 16.2hh, but some males may reach 17hh or more.

PERCHERON (FRANCE)

The Percheron comes from La Perche in Normandy in northern France. The breed is an ancient one dating back to 732, when Arab horses, abandoned by the Saracens after their defeat at the Battle of Poitiers, were allowed to breed with the local heavy mares of the region. From these matings the Percheron type emerged.

At this time the horse was much lighter than its modern counterpart and was used for riding as well as for light draught work. The type remained popular until the Middle

The Percheron is an ancient breed from La Perche in northern France.

Ages and the Crusades, when Arab and Barb horses from the Holy Land were mated with them. It was also around this time that the Comte de Perche brought back Spanish horses from his forays in Spain; these were also mated with the Percheron, with further infusions of Andalusian added later. By the 18th century, the original breed had become almost completely eradicated by the addition of Thoroughbred and more Arab; in 1820 two grey Arab stallions were mated with Percheron mares, which is responsible for the predominantly grey colour of the modern-day breed.

By now all the heaviness of the ancient breed had disappeared; consequently, heavy mares from other regions were bred with Percheron stallions to make them more suitable for agriculture and to formulate the breed as it is seen today. The lighter Percheron still exists and is used as a heavy riding horse, while the heavy version is still used for farm and forestry work and, in some countries, for pulling drays. It is also popular in the show ring.

Over the years the Percheron has been heavily exported to other

countries such as the United Kingdom, Canada, Australia and other parts of Europe, which has helped in its recognization as one of the world's leading heavy breeds.

The Percheron possesses a good deal of elegance due to large amounts of Arab blood which have been added over the centuries. It has an excellent temperament, is calm, obedient and easy to handle, and has a keen intelligence. It has a smooth but lively action, making it comfortable to ride. Mainly grey but occasionally black and dark chestnut. The small Percheron stands at 14.1–16.1hh, the large Percheron being 16.1–17.3hh.

SHIRE

The Shire is one of the most famous and distinctive of all the draught horses and one of the largest and most majestic breeds in the world. Descended from medieval warhorses, whose immense strength enabled them to carry knights into battle wearing full armour, it was probably based on the Friesian with later infusions of Brabant. It was brought to England by the Dutch to drain the fens of East Anglia. However, it was not until the late 19th century that the best heavy horses in England were selected to develop the breed as it is known today.

The Shire's strength also made it suitable for agriculture and heavy haulage work, so initially the breed was established in Lincolnshire and Cambridgeshire where strong horses were required to cope with heavy fenland soil; but the Shire soon became widespread in Staffordshire,

OPPOSITE & BELOW: The Shire is the most easily recognizable of the heavy breeds. It is descended from medieval warhorses.

Leicestershire and Derbyshire until it eventually spread over the whole of England.

Up until the 1930s, the Shire was widely seen across the country, but as farms began to substitute tractors the numbers dropped dramatically until the breed was in danger of disappearing altogether. Fortunately, the problem was realized by a few dedicated breeders who helped to promote the breed and restore its popularity.

The Shire Horse Society has worked tirelessly to raise funds and encourage the spread of the breed to other countries. Today there are active societies across Europe, the United States, Canada and Australia. Although a few Shires are still used on farms today, they are kept mainly for the sheer pleasure of working them in their traditional roles. They are also used in ploughing competitions, again, for pleasure, and for the same reason breweries use them in pairs to deliver beer locally.

The Shire's most significant feature is its sheer size and massive muscular conformation. It is the largest and strongest horse in the world and, when mature, weighs a ton or more. Built ultimately for strength, the chest is wide, the back short-coupled, the loins and quarters massive. The legs, joints and feet are sufficiently large to balance and support the Shire's size; the lower legs are covered with long, straight, silky feathers. Even though the Shire is such a large horse, it is not an ungainly heavyweight; in fact it is very much in proportion and quite beautiful to behold. The head is always noble and the nose slightly Roman. The eyes are large and wise. The Shire is well-known for its patient, gentle and placid nature; it is a true 'gentle giant', and is legendary for its kindness. It is the foundation breed for the Katahdin and Royal White breeds.

Bay, black, brown and grey are the recognized colours of the breed. White feathers on the legs are preferred for the show ring and white face markings are common. The Shire stands at 16.2–18hh.

SUFFOLK PUNCH

The Suffolk Punch originated in East Anglia in England and takes its name from the county of Suffolk, 'Punch' being an old word for short and thickset. It is thought to date back to 1506 and is the oldest heavy breed in Britain.

The breed was first developed by crossing the native heavy mares of the region with imported French Norman stallions. However, modern-day Suffolks can be traced back on the male side to a single, nameless stallion, foaled in 1768 and belonging to Thomas Crisp of Orford, near Woodbridge, Suffolk; even though the breed is relatively pure, infusions of Norfolk Trotter, Thoroughbred and cob were added during the centuries that followed.

The Suffolk Punch is immensely strong, but due to its relatively small size is also quite agile. These qualities, combined with a lack of feather on the legs, like the Percheron, made it ideal for working the heavy clay soils of East Anglia. As with many of the heavy breeds, numbers fell dangerously low when farm tractors became widespread. Today Suffolks are rare, even though there has been a concerted effort in recent years to increase numbers. Today, Suffolks are shown, used in ploughing competitions, or are owned by breweries.

The Suffolk Punch is always chestnut in colour (the traditional spelling for this particular breed is chesnut, without the 't'). The breed is well-known for being easy to train, and it is docile and hardworking. It is also capable of doing almost any kind of work and is easy to maintain. The Suffolk Punch stands at 16.1–17.1hh.

The strong but agile Suffolk Punch is a rare breed from East Anglia in England. It is invariably a rich chestnut colour.

INDEX

ACKNOWLEDGEMENTS

Photographic Acknowledgements

Front Cover: Main image–Flickr Creative Commons/Nigel Webb. Insert left–©istockphoto/Iain Sargeant. Insert right–©istockphoto/Loretta Hostettler. Spine–©istockphoto/rtyree1. Back Cover: ©istockphoto//TSnowimages.

The following photographs were supplied courtesey of the American International Marchigiana Society: pages 83, 84.

The following photograph was supplied by the American Karakul Sheep Registry: page 208.

The following photographs were supplied through ©istockphoto, courtesy of the following photographers: Alan Crawford: page 235. Alkimson: page 141. AtWaG: pages 12 below, 13. Cathleen Abers-Kimball: page 81. Chris Elwell: pages 2, 15, 233. Chris Hepburn: pages 137, 140. Chris Pole: page 48. Craig Walsh: pages 3, 120. Daniel Gustavsson: page 45. Daniel MAR: page 149. Daniel Sainthorant: page 96. Dave Rankine: page 215. Elmholk: pages 64, 67. Eric Coia: page 218. Ernst Kempf: page 59. Eugenio Barzanti: page 47. Frances Twitty: page 90. Geno Sajko: page 209. Gregory Bergman: page 72. Hanne Melbye-Hansen: page 116 below. Iain Sargeant: page 139. James Harrop: pages 189, 190. James Whittaker: pages 77, 79. Jeroen Peys: page 6 below. Jocrebbin: page 57. Jonathan Wong: page 18. Judy McPhail: pages 144 above, 157. Karen Low Phillips: page 160. Kawisign: page 20. Iahulbak: page 87. La Nae Luttrell: page 69. Leslie Banks: page 29. Lidian Neeleman: page 60. Linda & Colin McKie: page 34. Lisa F Young: page 32. Loretta Hostettler: page 214. Margo Harrison: pages 24, 27. Markus Drach: page 237. Melissa Carroll: page 36. Michael Westhoff: page 205. Michele Vacchiano: page 145. Mike Bentley: page 12 above. Morten Normann Almeland: page 51. Mr_Jamsey: page 146. Nancy Nehring: pages 100, 227. Paulina Lenting-Smulders: pages 30, 169. Pawel Cebo: page 144 below. Perry Watson: page 153. Peskey Monkey: page 17. Pete Pfändler: page 102. Peter Elvidge: page 63. Petrus van der Westhuizen: page 194. Robert Asento: page 10. Rtyree: page 162. Rupert Kirby: page 206. Sally Wallis: page 207. Shannon Forehand: page 109. Sheryl Griffin: page 166. Stefan Ekernas: page 35. Suemack: page 226. Terry Swartz: page 142. Thomas Bradford: pages 8-9. Thomas Dickson: page 71. TSnowimages: pages 4-5. Verena Matthew: page 26. Vika Valter: page 198. Ymgerman: pages 111, 112 below, 113.

The following photograph were supplied through Flickr/Creative Commons www.creativecommons.org courtesy of the following photographers: Acradenia: page 80. Amander Slater: page 130. Arpingstone: page 11. Bellybe¬_Khe: page 116 above. Clarity: page 164. Cliff1066TM: page 172. Dave Hamster: page 128. Dave Merrett: page 122, 123. Goat.Pirate: page 165. Gumdropgas: page 247. Hans Splinter: page 54. Jennifer Schwalm: page 153 above. Jon Stammers: page 170. Jos: page 55 above. Juliaabrown: page 7. Julz91: page 133. Just Chaos: pages 125, 159, 213. Magic Foundry: page 177. Mozzercork: page 76. Pascal: page 106. P Markham: page 168. Spencers Brook Farm: page 118. Redjar: page 74. Richard Hawley: pages 250, 251. Robert Scarth: page 14. Robert Son of Randy: page 181. The-Gut: page 245. Twicepix: page 40. Wolfiewolf: page 178. Woodleywonderworks: page 121.

The following photographs were supplied through Wikimedia Commons courtesy of the following photographers: 20100: page 107. A'chachaileith: page 219. Aleks: page 91. Alex Erde: page 242. Amanda Slater: page 124, 131. Ancalagon: page 22. Andre Bianco: page 89. Andrew: page 230. Annie Kavanagh: pages 52, 53. Arjecahn: page 235. Aracademia: page 212. Asa Rosenberg: page 143. Berteun: page 55 below. Bill Tyne: page 171. Bob Nichols: pages 186, 187. Böhringer Friedrich: pages 16, 25, 41, 152. Carly & Art: pages 132, 238, 239. Caroline Ford: page 138. Catriona Savage: page 229. Cgoodwin: pages 85, 94, 98. Chad K: page 174. Charles Drake: pages 104, 105. Cooper: page 38. Coopertje: page 201. Crackercattle.org: page 56. Cyrille Bernizet: page 82. Danielle Langlois: page 161 above. Darren Wyn Rees: page 114. Dave Merrett: page 119. Dave Pape: page 21. Davide Ferro: page 154. Dirk Ingo Franke: page 59 above. Deviers.fabien: page 210. Don't Worry: page 134. Dorothea Witter-Rieder: pages 220, 221. Drew Avery: page 127. Drichard: pages 174 above, 188. Ed Schipul: page 110. Eric Peterson: page 148. Eweri: page 197. Félix Potuit: page 43. Fernando Hartwig: page 112 above. Fir0002/Flagstaffotos: page 68. Garitzko: page 115. Ggonell: page 135. Gilbert Le Moigne: page 78 above. Glenn Brunette: page 211. Grimlock: page 86 above. Hedwig von Ebbel: page 58. James Thompson: page 103. Jean-Pol Grandmont: page 97. Jennifer Dickert: page 150. Jenny: page 200. Jjron: pages 6 above, 66. Jude: page 126. Just Chaos: pages 158, 173, 175, 216, 217. Kersti Nebelsiek: page 33. Kitkatcrazy: page 228. Kreg Leymaster: page 193. LesMeloures: page 246. Lisbeth Landstrøm: page 231. Liz Lawley: page 156. Ltshears: pages 147, 161 below, 163, 167. Man Vyi: page 75. Markus Braun: page 78 below. Marjon Kruik: pages 244, 249. MichaelW GFDL: page 61. Michel d'Auge: page 86 below. Monica: page 46. Montanabw: page 253. Moriori: page 151. Paul: page 176. Poinciana: page 95. Robert Scarth: pages 31, 101. Robin Lucas: page 243. Rs-foto: page 241. Rudi Riet: page 225. Ruestz: page 37. Sarahemcc: page 19. Saruman: page 182. Sean/Someone's house pet: page 108. Smurrinchester: page 44 left. Son of Groucho: page 49. Soxophone Player: page 184. Stephen Ausmus: page 203. Steve & Jem Copley: page 23. Steve Walling: pages 224, 234. Teddy Ilovet: page 195. Thomas Quine: page 204. Thor Rune: page 88. Tswgb: page 62. Wikimedia: page 117. Xabiercid: pages 179, 180. Xocolatl: page 39.